Big Steel

Daniel Madar

Big Steel
Technology, Trade, and Survival
in a Global Market

UBCPress · Vancouver · Toronto

20 19 18 17 16 15 14 13 12 11 10 09   5 4 3 2 1

Printed in Canada with vegetable-based inks on FSC-certified ancient-forest-free paper (100 percent post-consumer recycled) that is processed chlorine- and acid-free.

Printed in Canada on acid-free paper

Library and Archives Canada Cataloguing in Publication

Madar, Daniel R., 1941-
  Big steel : technology, trade, and survival in a global market / by Daniel Madar.

Includes bibliographical references and index.
ISBN 978-0-7748-1665-6

  1. Steel industry and trade. I.Title.

HD9510.5.M33 2009   338.4'7669142   C2008-907950-7

Canadä

UBC Press gratefully acknowledges the financial support for our publishing program of the Government of Canada through the Book Publishing Industry Development Program (BPIDP), and of the Canada Council for the Arts, and the British Columbia Arts Council.

This book has been published with the help of a grant from the Canadian Federation for the Humanities and Social Sciences, through the Aid to Scholarly Publications Programme, using funds provided by the Social Sciences and Humanities Research Council of Canada.

UBC Press
The University of British Columbia
2029 West Mall
Vancouver, BC V6T 1Z2
604-822-5959 / Fax: 604-822-6083
www.ubcpress.ca

Contents

Preface / vii

Acknowledgments / ix

1 Introduction / 1

2 A Tough Industry / 17

3 Prices, Preferences, and Strategy / 61

4 Trading Steel / 83

5 Survival / 133

6 Steel in a Global Perspective / 175

Notes / 195

Bibliography / 223

Index / 234

Preface

Positioned at the centre of manufacturing, the steel industry is a key economic sector. Its member companies share changeable financial health, exposure to strongly cyclical demand, vulnerability to oversupply, and a tendency toward price warfare. The most common grades of steel are highly standardized, readily traded, and widely usable. Trade over great distances is encouraged by economies of scale and favourable transportation costs, enabling buyers to enjoy diverse supplies and competitive prices. World steel production grew slowly until this decade, when demand in Asia began to soar. In response, output increased by 58 percent, reaching 1.3 billion tonnes in 2007. More than one-third of that is exported, and 81 percent of net exports come from China, Japan, Ukraine, Russia, and Brazil.

With that prodigious volume, misalignment of production and demand can be severely problematic. When there is excess supply, the industry's economics tempt producers to cut prices instead of output. That encourages price warfare, which is a dangerous game in an industry with high fixed costs. Trade makes it possible for overstocks in home markets to be shifted elsewhere and, when local prices require it, to be offered at discount. Prices themselves vary widely – a 337 percent increase in this decade for hot-rolled coil, for example – and economic downturns can lead just as quickly in the opposite direction. All of this makes the industry a contingent milieu, and with most forms of trade protection illegal under WTO rules, producers are on their own. Major steelmakers are consolidating as they search for stability and diversified markets. The unexpected merger in 2006 of the world's two largest, Arcelor and Mittal, portends a massive global consolidation. Soon after that event, Canada's three big producers were acquired by steelmakers from Europe, the United States, and India.

Canada and the United States are each other's largest steel suppliers, and the industry's multiple products flow in both directions. Why that trade exists in an active and competitive world market can be explained by proximity, as the two countries' steelmakers are located around the Great Lakes heartland and serve the same industries. Familiarity and common commercial networks are additional advantages. Gaining those advantages is one reason why foreign steelmakers – including Brazilian, Indian, and Russian ones that were once on the industrial world's periphery – acquire Canadian and American firms.

The steel industry's position in the international economy is the subject of this book. Discussed are the economics and technology of steelmaking, pricing and export strategies, the industry's migration to Asia and Latin America, survival strategies, and the industry's future in a globalized economy. Living not too far from the steel town of Hamilton, Ontario, has made me aware of the industry's sheer size and presence. Imagining tall furnaces, massive mills, ships unloading tonnes of coal and ore, and plumes of flame illuminating the night should evoke in the reader a suitably expansive sense of this book's subject. One of its recurring themes will be scale economies, which is a concept fully appropriate to this monumental industrial entity. An understanding of the situation and workings of steelmakers is this book's objective.

Acknowledgments

I would like to thank my colleagues Lewis Soroka for reading part of Chapter 2 and Livianna Tossutti for checking my statistics. Appreciation goes also to the anonymous reviewers for their consideration of the book in its intended breadth and for their helpful and practical suggestions. Responsibility for any errors or flaws that remain is, of course, mine. Most of all I would like to thank my UBC Press editors: Emily Andrew for her encouragement and advice as the project progressed, and Randy Schmidt for ushering it expeditiously into print. My appreciation goes to the production staff at UBC Press for their promptness, graciousness, and professionalism and especially to my production editor, Megan Brand, for her prompt and enthusiastic management of turning a manuscript into print. Finally, I am grateful to the International Iron and Steel Institute for permission to use the dramatic graph of world steel production in Chapter 1 and to Corus Steel for permission to reproduce the four diagrammatic figures of steel production in Chapter 2.

1
Introduction

The invention of the Bessemer converter in 1855 made it possible to produce steel in volume, transforming an expensive specialty product into "the quintessential material input" of modern economies.[1] Steel possesses an ideal combination of versatility and great strength. "Upon impact it does not break, shatter or easily distort ... and it can be rolled into shapes and subjected to temporary or continuous tensions without its ability to perform being seriously affected."[2] Modern steelmaking technology can produce enormous quantities in a wide range of products and grades, and improvements in process efficiency have made the metal relatively inexpensive. Steel is the mainstay of some of the biggest industries – transportation, petroleum, machinery, shipbuilding, appliances, and construction – and the volumes involved make the steel commodity market, at $1 trillion annually, the world's largest.[3] Steel can be re-melted, making it also the world's most recycled product.

In 2004, world steel output passed the milestone of 1 billion tonnes – up from 750 million tonnes only eight years previously – driven by burgeoning growth in China and India. World steel production for 2007 was 1.3 billion tonnes. Figure 1.1 shows the pattern in tonnes produced. In rates of growth, there was actually an 0.5 percent decline between 1990 and 1995, and growth of 2.4 percent between 1995 and 2000. Then came the acceleration: 6.2 percent between 2000 and 2005 and 8.3 percent in the two years alone between 2005 and 2007. By comparison, the rate of growth between 1970 and 1990 averaged 1.3 percent.[4]

Turbulence and instability characterize the industry, and the purpose of this study is to understand its situation in an international economy in which steelmakers have proliferated, governments have withdrawn their historic supports and protections, and steel has become a widely traded commodity. For established producers in the industry's historical homeland of Europe and North America, the past four decades have seen a stark transformation.

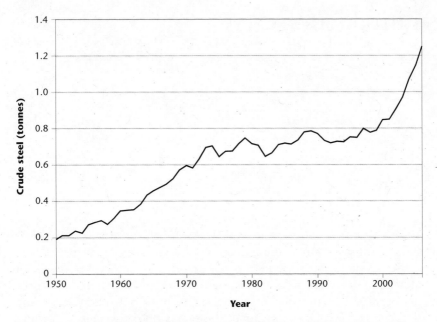

Figure 1.1   World crude steel production, 1950-2007. *Reproduced with permission of the International Iron and Steel Institute.*

In Europe, struggling steelmakers in both the state and the private sector have been consolidated and privatized, with the most viable parts combined into new firms and governments relieved of expensive and often obsolescent enterprises. One result has been Arcelor, an amalgamation of producers in France, Spain, and Luxembourg. In the United States, where there had been no state ownership, the process proceeded through bankruptcies, with rescue then coming from investors and consolidators. The largest successor was International Steel Group, whose components included the former Bethlehem Steel. There has been some of each measure in Canada, with the Ontario government assisting the recovery of one producer, Algoma Steel, while leaving the other, Stelco, to face bankruptcy court.[5] At the same time, Canada has been spared the United States' and Europe's large mill closings. Government protection against import competition was withdrawn as members of the General Agreement on Tariffs and Trade (GATT) and subsequently the World Trade Organization (WTO) agreed to reduce tariffs and other trade barriers in a series of agreements beginning in 1948. These changes have left European and North American producers to their own devices in fully con-testable national markets.

The state's withdrawal of patronage compels us to focus on the industry itself. Here, the global trade in steel directs our attention toward pricing, market-capturing strategies, and the industry's international dispersion. At the heart of the industry's workings are technologies and cost structures, which make dramatic turns of fortune an abiding prospect and survival a mutual concern. Across the industry's global expanse, consolidation is now blending once-national producers into entities that are new, large, and cosmopolitan – a dramatic result of steel production's migration to new locales.

While the industry in Europe and the United States was reaching its full maturity in the 1960s and beginning to pass into obsolescence (Canada's two main producers, Dofasco and Stelco, managed to avoid that decline – Stelco's bankruptcy came later, in 2004), new producers were emerging in Asia and Latin America. Though they were originally established to supply expanding domestic manufacturers, they were soon exporting their surpluses and beginning to treat foreign markets as key ones. Their efforts generated trade complaints in Europe and North America and partial protection under anti-dumping rules. Since then, privatized Russian steelmakers, facing limited home demand, have turned to exporting as well. More dramatically, China has suddenly emerged as the world's largest producer. The volume involved – that country accounts for fully one-third of world production – has raised concerns that Chinese steelmakers' determined quest for growth and revenue could overshoot demand, flood world markets with surplus steel, and collapse prices. The industry's cost structure has always made episodes of low prices dangerous for weaker producers; the prodigious capacity of China's steelmakers now worries even the strongest ones.

New producers have become consolidators, and steelmakers in Europe and North America, having shed inefficient capacity and invested heavily in new equipment, have become attractive takeover targets. By buying Inland Steel and later International Steel Group, Mittal Steel became, along with United States Steel and Nucor, one of North America's three largest producers. Mittal Steel had itself grown to global dominance through consolidation. Its founder, Lakshmi Mittal, had begun by acquiring mills in Asia, Mexico, and the Caribbean and then expanded to become the world's second-largest producer. In 2006, Mittal stunned the industry by buying the world's largest producer, Arcelor.[6] The new firm is massive, with output triple that of Nippon Steel, the next-largest. The merger raises the unprecedented prospect of the industry consolidating globally as major producers acquire smaller ones and, as did Mittal and Arcelor, combining with one another.

Consolidation is also occurring in other commodity metal industries. In 2006 Brazil's Vale (formerly Companhia Vale do Rio Doce, or CVRD), the world's largest iron-ore producer, purchased the world's second-largest nickel producer, International Nickel Company; and in 2007 the aluminum producer Alcan Inc. was bought by the British-Australian mining conglomerate Rio Tinto Group in a deal representing one of the largest foreign takeovers in Canadian history. For both acquiring companies, the goal was to supply the rapidly growing demand for industrial metals in China and India.

In the steel industry, one reason for consolidating in the wake of the Arcelor–Mittal merger is the defensive one of assembling comparable capacity. A similarly defensive reason is to grow large enough to close facilities in lean years and still remain in business. Closely related is the desire to establish a presence in multiple markets in the hope that downturns in some would be offset by stronger conditions in others. Gaining access by buying incumbent producers is yet another reason. Still another is achieving complementarity. For producers of high-value–added steels, that involves acquiring suppliers of semi-finished steel. For semi-finished steel producers, conversely, that means acquiring advanced processing facilities. A final reason is to secure proprietary technology and long-term clients – important assets in the lucrative and competitive top end of the market. In that segment, a growing demand for thin, strong, and formable sheet steel comes from the world's automobile industry as it seeks to produce lighter yet more crash-resistant vehicles. Mastery of sophisticated metallurgy made firms such as Arcelor quite valuable.

Consolidation proceeds apace. Canada's Dofasco was purchased by Arcelor in 2006 after a bidding war with Germany's ThyssenKrupp AG, and Algoma Steel a year later by Essar Global Ltd., a large, Mumbai-based conglomerate whose steel subsidiary is India's top flat-steel exporter. In July 2007, Stelco, having emerged from bankruptcy, announced that it was for sale; the following month it was purchased by United States Steel Corporation. The three firms are now known, respectively, as ArcelorMittal Dofasco, Essar Steel Algoma, and United States Steel Canada. On the American side of the border, in 2004 the Russian steelmaker OAO Severstal purchased Rouge Industries, the Ford Motor Company's spun-off steelmaker, and then made a series of subsequent acquisitions to become the fourth-largest American producer. What made these purchases attractive – even that of the decrepit Rouge Industries – was their position in North America's automobile industry. Enhancing Stelco's appeal was its capacity to supply steel for processing into high-margin products.

Larger firms are also prime candidates. United States Steel, whose output is four times that of either pre-merger Dofasco or Stelco, is still medium-sized compared to ArcelorMittal. Making the company a potential takeover candidate is its profitability and position in the American market. Corus Steel's similar position in the European market – where it was the second-largest producer – led to its purchase in 2007 by India's Tata Steel. As one index of the industry's structural transformation, a major component of Corus was the privatized remnants of British Steel.

Industry and the Position of the State

As this sketch shows, the steel industry epitomizes globalization, which can be defined as the spread of production, markets, and investment across national borders. Globalization is the result of the ending of state controls on the movement of goods and capital. Successive rounds of trade liberalization under GATT and its successor, the WTO, have reduced or eliminated tariffs and bound members to the non-discriminatory treatment of foreign goods. By agreeing to tariff reductions, states have given up their power to restrict imports by taxing them. More generally, under the WTO principle of national treatment, states are expected to impose no special disadvantages on foreign goods or special favours on domestic ones. These measures have opened the way to a massive increase in global trade. Governments, eager to promote exports, have added their encouragement.

Steel is readily exported because it is highly standardized, particularly when it comes to commodity-grade products such as structural steel, coiled sheet, and steel plate. To take a common example, one mill's steel reinforcing bars are direct substitutes for another's. For any good, product differentiation limits potential buyers and restricts trading opportunities; in steel, however, such differentiation affects only top-of-the-line specialty grades. For the much broader array of commodity-steel products, qualitative barriers are very low. The key differentiation is price – a fact with significant trade implications, as will be seen shortly. Overall, 36 percent of world steel output is exported – up from 22 percent in 1975 – and products circulate widely.[7]

This commerce reflects a fundamental reorientation. Historically, steel producers served their own national markets and occupied their respective positions within national industries. States regarded steel as a strategic commodity. Its importance in manufacturing – especially in armaments production – made it too vital to be entrusted to foreigners and too vulnerable to wartime interruption. In national economies more generally, steel represented massive concentrations of investment, employed thousands of workers, and was at the core of key industrial regions. Some governments nationalized the industry; others encouraged or tolerated cartels. Since then, as states have withdrawn their custody, the basis of ownership has evolved from national to global. Privatization has exposed steel producers to the financial markets, and the removal of investment barriers under WTO, regional, and bilateral agreements has made those producers available to foreign buyers.[8]

Governments might still wish to keep their steel producers under domestic control, but their power to do so is limited. That much was evident in the French government's vigorous efforts to block Mittal's takeover of Arcelor. The government regarded Arcelor as a national champion and worried about mill closings and layoffs. Though its efforts, joined by those of Arcelor's management, delayed the sale, it did eventually proceed. Appeals to the European Union (EU) were to no avail. As an indicator of openness, however, the Arcelor case is exceptional in that other governments have not resisted international steel takeovers, though they have been attentive to the effects on competition. The biggest exception is China, which limits foreign participation to joint ventures, though conversely the government regards foreign takeovers as a suitable way for Baosteel, its major producer, to expand.

Nonetheless, in 2007 large foreign takeovers began to raise concerns in both Canada and the United States. In June 2007, Canada's Minister of Finance announced the formation of a Competition Policy Review Panel (CPRP) to examine Canada's Competition Act and the Investment Canada Act, with particular attention to whether the legislation should be changed to address investment by state-owned entities and considerations of national security.[9] This reconsideration was prompted by public reaction to a spate of very large foreign takeovers, including Arcelor's acquisition of Dofasco.[10] In October 2007, the industry minister ordered that issues concerning investment by state-owned enterprises and their national security implications be removed from the panel's mandate – an indication that the government might take up the question separately. A key recommendation of the panel's report, issued on June 26, 2008, was that existing investment restrictions should be reduced in the interest of improving productivity.[11] In July 2007, President George W. Bush signed the Foreign Investment and National Security Act. From now on, the treasury department's Committee on Foreign Investment, which monitors foreign acquisitions, will be required to fully investigate those investments that involve companies owned or backed by foreign governments.[12] Until then there had been no broad restrictions on foreign investment except in sectors where there was a risk to national security. The new American law was taken as a model by the German government, which was concerned that existing German and EU rules were not specific enough to cover foreign-government–backed entities, particularly Russian and Chinese ones.[13] Legislation passed by Germany to amend the Foreign Business Act now allows the government to block foreign investment of more than 25 percent in a German firm. The legislation was modified in January 2008 at the request of the European Commission (EC) to exempt firms of the European Union.[14] Meanwhile, in Japan, business reacted with anxiety to a government plan to change merger and acquisition rules so that the subsidiaries of foreign firms would be able to pay for takeovers with the parent firm's shares. The worry was that Japanese firms would be more vulnerable to large and wealthy foreign buyers. Specifically mentioned were Japanese specialty-steel producers.[15]

For those doing the buying, steel acquisitions are attractive both domestically and internationally. A strong domestic attraction is access to the acquired producer's home markets. A strong international attraction, created by liberalized trade, is the prospect of assigning facilities to entire world markets. Doing so can involve simply exporting items currently in production or, more expansively, investing in a fully specialized mill. The main constraints, representing costs of trade, are locational and logistical. The fact that those costs are much lower in domestic than in international trade suggests why buying access to a producer's home market is a popular strategy. If the cost difference were to disappear, very different ways of organizing world steel production would become feasible.[16] At the same time, the amount of steel traded internationally, as was just seen, is substantial – an indication that foreign markets are served quite readily at current costs of trade.

Another limit on state authority is WTO rules that forbid the subsidization of exports. Subsidies paid directly to producers are permitted; however, these are actionable when they cause harm to the producers of other WTO members, when they impair the common benefits of WTO rules, or, more generally, when they cause "serious prejudice to the interests of another WTO member."[17] If those effects can be established, subsidies are subject to countervailing tariffs. WTO rules encourage members to resolve subsidy disputes through consultation. With steel products, such consultation occurred in 2002 when the EU, in multilateral discussions under the auspices of the Organisation for Economic Co-operation and Development (OECD) about the health and future of the steel industry, agreed to eliminate almost all subsidies to its steelmakers. In formal WTO proceedings, most cases have involved export subsidies. Much more infrequent are cases involving non-prohibited but actionable subsidies because they must be shown specifically to cause injury. Such harm is difficult to prove because the trade effects of payments to producers, in contrast to export subsidies, are usually more diffuse. The result has been a general reluctance to challenge other members' domestic industrial practices.[18] An exception is the United States, which uses direct industry subsidies less than its trading partners and has less reason to fear "retaliatory investigations."[19] Because actionable subsidies may be challenged by any WTO member, however, states awarding them are aware that they do not have an entirely free hand.

Anti-dumping complaints are easier to prove and less expensive to litigate, and through them states have retained a powerful tool of intervention. Under WTO standards, states may impose tariffs to eliminate discount margins on imports. Investigation must show that imports are being sold below home-market prices or below the cost of production. If material injury is found, complainants can receive tariff relief. Safeguard rules are a similar but less potent measure. They allow the use of tariffs to remedy injury caused by an import surge; and unlike anti-dumping rules, they do not involve allegations of unfair practice and do not single out particular exporters. However, safeguard actions under current WTO rules are governed by stringent requirements and may be disallowed on appeal to the WTO.[20] Failure to comply opens the way to retaliation by affected trading partners. That prospect forced the Bush administration to revoke a set of special steel tariffs imposed under safeguard provisions in 2002. In contrast to regular tariffs, which have steady effects, both anti-dumping and safeguard interventions are intended to operate exceptionally and case-by-case in very specific product subcategories. For producers facing longer-term international competition, neither is a dependable protection.

A third tool of intervention is competition law. By making collusion illegal and by authorizing measures to prevent or disband monopolies, it provides a means – should states choose to use it – of regulating the effects of consolidation. This bears on one incentive for global steel mergers – maintaining price stability by coordinating output levels among a workably small family of producers. The incentive to coordinate is rooted in the steel industry's cost structure, as will be seen shortly.

The world's automobile industry provides a preview of the kind of integrated global steel market made possible by trade liberalization. With an annual output of some 50 million cars whose average steel content, by weight, is 70 percent, the automobile industry is a prime customer. Consolidation has occurred there as well, with smaller firms such as Saab being acquired by larger ones such as General Motors (GM). With that has come the sharing of components and vehicle platforms. Product standardization allows component production to be dispersed internationally and assembly plants to be located in the large markets of Europe, the Americas, and Asia. For steelmakers, this opens an expansive prospect. Although North America's Big Three automakers descended into grave financial difficulty as a result of the global financial crisis of 2008, the longer-term outlook is that world automobile production will continue (although the automobile industry may be a re-organized one). Demand for automotive steels will also continue.

Though automakers prefer to deal with nearby mills, standardized requirements and dispersed production enable steelmakers to become global providers. One way of doing so is by acquiring steel firms, such as Dofasco and Stelco, which have an established automotive business. A more comprehensive strategy involves helping design vehicles that require a steelmaker's proprietary sheet steel and that are slated for manufacture at multiple world sites. With that eventuality in mind, major steelmakers are entering into increasingly close product-development relationships with world automobile producers. The best-positioned candidates are those, such as ArcelorMittal, that have design expertise and that can supply their proprietary product worldwide either on their own or in co-operation with other steelmakers.

Sectors, Cost Structures, and Prices

It is easy to view industries as more or less alike. Trade politics studies, for example, often assume that firms demanding protection are motivated by simple rent seeking. That assumption makes it possible to model firms' behaviour mathematically and to portray purposeful political advocacy, but it too readily generalizes among industries that may be very different.[21] Modern economies have multiple sectors with differing capital structures, technologies, markets, and international exposure. That fact became strikingly clear to researchers investigating the applicability of strategic trade policy. After studying the telecommunications, jet transport aircraft, automobile, and steel industries, they concluded: "The features of the sectors we discussed were so strikingly different, when examined with a sufficient degree of care, and the problems confronted by firms in those sectors were so disparate, that it made little sense to speak of a 'trade policy problem.'" This was true in industries that were themselves conveniently regarded in aggregate: "even across areas that might be thought to have much in common as a class – 'high-tech' or 'smokestack' industries – careful examination of their market structure, conduct, and performance demonstrated clear and particular differences." The trade problems of the automobile industry, the researchers found, were quite different from those of their smokestack industry counterparts in steel. For trade policy, also readily characterized in aggregate, the clear implication is that different industries require different treatments.[22]

Steel production shows how dissimilar different parts of the same industry can be. Technology divides the steel industry into two very distinct sectors – integrated producers and minimills. Integrated producers (a shorthand reference to vertical integration) refine iron ore into steel and roll it into an array of products. The process is continuous, requiring close coordination among adjoining facilities. The volumes involved make those facilities massive. Those facilities in turn impose formidable capital requirements and expensive cost structures. Minimills roll steel products from melted scrap. Their comparatively simple and cheap technology gives them a cost and price advantage, and they are efficient at moderate volumes. In an undifferentiated industry, firms are affected by the same conditions in the same way. In the steel industry, however, conditions related to capital costs and minimum efficient scale have created problems for integrated producers and opportunities for minimills. Because the integrated sector is the one that has been beset by difficulties, and because it accounts for two-thirds of steel production worldwide, it is the focus of this study. Integrated producers will be referred to simply as steelmakers unless minimills are being mentioned specifically. For the same convenience, steelmakers – again, unless minimills are a point of focus – will be spoken of as the steel industry.

To proceed beyond categorical assumptions, it is necessary to understand the steelmakers' basic technology and economics. The key fact to appreciate is that cost structures and cyclical demand make the steel industry inherently volatile and difficult. Expensive technology imposes high fixed costs that must be covered through good times and bad. Because that technology is efficient at high volumes and produces in a continuous process, it encourages producers to cover fixed costs by maintaining output. They can do so in the short run because variable and marginal costs are normally comparatively low. Yet the demand for steel derives directly from the demand for durable goods, and that demand varies with the business cycle. When steel demand falls, the combination of fixed, variable, and marginal costs makes it reasonable to cut prices instead of output. When all producers do the same, they depress prices further, and those unable to persist, given the high costs they bear, face bankruptcy. Cutting output is an option when variable costs are high, but that condition does not often occur in integrated steelmaking. Steel's substitutability makes prices pivotal, and high volumes make small differentials important. In the same way, steelmaking's standardized and portable technology makes the same efficiencies and volume-based pricing power available to anyone who uses it. Though that technology reduces labour-cost differentials between producers in wealthy industrial and industrializing countries, it also makes direct competitors of producers in places such as China.

Capital-cost barriers and scale economies in steelmaking have always provided the structural basis of oligopoly, and vulnerability to overproduction and low prices has provided incentives for steelmakers to actually behave as an oligopoly. Competition law forbids price fixing, but for a long time a system of price leadership was possible in the American steel industry, and in Canada an informal specialization in particular submarkets limited direct rivalry. Stability broke down in the 1960s when minimills, which had strong cost advantages and no interest in coordinating prices, began entering the industry and when national steel markets became open to imports. Minimills were reluctantly accommodated, but imports were challenged in trade tribunals. Minimills and imports contributed directly to the industry's travails and forced its reorganization. An exacerbating factor was that many of the imports came from new producers using the latest technology. That drove home the fact that the industry's large and segmented production facilities, together with its weighty cost structure, make modernization piecemeal and expensive, particularly for producers most in need of it and especially when prices are under pressure.

Cooperative incentives have reappeared at the global level. The scale of the largest steelmakers, along with accumulating surplus capacity in the industry, raises the same worry that once beset national markets: producers seeking to increase their own income will lower prices for everyone. National markets, in which supply and demand could be kept aligned, are no longer a protection. In an open trading system, surpluses in one country can be shifted to others. As was once true of producers in national markets, however, globally consolidated producers have both the limited numbers and the shared incentives to develop a common understanding of production levels and prices. Lakshmi Mittal himself has expressed hope for this result.[23]

Prices are indeed the central element. As the point of differentiation in a highly standardized array of products, they figure centrally in strategies to capture domestic and foreign market share. Even more practically, prices mediate success and failure. Understanding prices, and the incentives and constraints that are derived from them, requires a basic knowledge of steel-making technology and cost structures. In the same way, understanding the setting in which pricing strategies operate requires attention to oligopoly.

Survival, Stability, and the Public Interest

For steelmakers in open economies such as Canada and the United States, what is at issue is survival and stability amidst global competition without the protection and patronage of the state. There are three basic options: protectionism, consolidation, and aligning in close relationships with major global customers such as the automobile industry. None of these is surefire. The most direct forms of protectionism are now illegal under WTO rules, and available expedients – most notably anti-dumping actions – operate partially and unpredictably. Consolidation offers the ability to achieve synergistic efficiencies among multiple installations, to serve multiple markets, and – assuming like-minded incentives among fellow producers – to achieve market stability. Competition authorities, however, have the power not only to prosecute price fixing but also to prevent market dominance by blocking mergers and ordering divestitures. Regulation may not be the only obstacle. Producers, even with sufficient latitude, may not be able to maintain a collective discipline. If they were to cut back their own output in the face of weak prices, they would have reason to fear creating market space for their rivals. More generally, the production and investment decisions that determine global capacity are guided by individual incentives with no common interest or strategy. One consequence is surplus capacity in addition to surplus production. The third option – aligning with major customers – promises the benefits of interdependence and shared commitments, but alignments can change. A steelmaker's partner has the buyer's free hand to choose and discard. As the suddenly dire circumstances of the Big Three automakers showed in 2008, partners can also fail. Since none of these options is fully secure, the steelmakers' future in considerable measure is what they and the market make of it. Under the conditions of globalization, that is an open prospect with all its attendant advantages and hazards.

Political economy has much to say about the incentives of economic actors and their connection to the public interest. An important balance of interests appears in the formal literature on trade protectionism, in which the theoretical efforts of the economist Jacob Viner continue to be influential. Anti-dumping actions, in his view, must reconcile a cheaper and reliable foreign supply, which lowers costs for the public, with the weakening or elimination of otherwise viable domestic producers, which represents a prospective loss to the public of both competitive prices and usefully invested capital. In making assessments, the focal interest is the public's and not the producer's. A similar emphasis can be seen in competition law, whose purpose is to serve the public interest by prohibiting producers from curtailing or eliminating competitive markets.

In the same way, a justification for anti-dumping rules is that they prevent producers, particularly ones enjoying a protected home market themselves, from gaining market share abroad by cutting prices long enough to drive out incumbent producers. Whether such predation actually occurs in trade has been a subject of empirical research, and trade economics has devoted much attention to protectionism more generally. A related question is this: What kinds of producers are the most likely to seek protection? An important consideration is factor mobility – being able to shift resources from one application to another. The steel industry's specialized and expensive equipment is not particularly flexible. Proof of this can be seen in the industry's closing of plants that were unable to meet lower prices and conceding entire product lines to minimills and imports. Another proof can be seen in the industry's historically frequent use of anti-dumping actions – a fact that originally prompted this study.

Why did steel production become established in Asia and Latin America, and why were its new firms able to challenge the incumbents? Current literature in economic geography has concise and powerful explanations for industrial location, the development and persistence of industrial regions, and the advent of trade. Trade advantages, the literature asserts, are not rooted immutably in particular places; they can be created by states. By sponsoring investment in new industries, states can lock in factors that would otherwise be unstable or temporary. Key to success are economies of scale and transportation costs. States in Asia and Latin America, acting on one view of the public interest, played an active role in promoting steel production. By helping these industries become established, they encouraged the migration of technology and altered the international division of labour. That was possible because steelmaking technology can be installed anywhere in the world and is for sale. The pattern of international steel production may change if energy costs, which rose sharply in 2007 and significantly increased the costs of transportation, were to return (a prospect considered in Chapter 6).

These states have withdrawn from direct control and ownership, and some steel producers, notably in Brazil, have done much better on their own. Industrial migration has entered a second stage as successful Asian and Latin American producers, joined by privatized Russian counterparts such as OAO Severstal, use their profits to acquire established firms in Europe and North America or enter into shared ventures. From a global perspective, these developments represent a reverse flow of industrial ownership and organization from periphery to centre. The results can be sizable. The second-largest minimill operator in North America is Gerdau Ameristeel, whose majority owner is Brazil's Gerdau SA. In 2007, it acquired Dallas-based Chapparal Steel, North America's second-largest structural steel producer. Although Gerdau, unlike Brazil's other steelmakers, was always in the private sector, it exemplifies the reverse pattern. Similar movement can be seen with Tata Steel's acquisition of Corus Steel and Essar's acquisition of Algoma Steel. All of these developments are further steps in the steel industry's global consolidation. They go well beyond the original motives of government sponsorship, and the link to the public interest becomes ever more indirect as these firms pursue autonomous interests of their own. Even so, the initiating condition, which combined state action with a public purpose, is worth remembering.

The Plan
As this brief survey illustrates, the steel industry's recent evolution suggests how changes that have encouraged globalization have affected one of the key sectors of modern economies. Steelmaking's situation at the centre of industry makes its condition vital. As has also been seen, these matters resolve into four topics: the technology and economics of steel production, approaches to contesting markets amidst oligopoly, the industry's international dispersion, and strategies for surviving in a global environment. A fifth topic that follows directly from these is the steel industry's future in a global economy, and the implications that follow for industries more generally.

Chapter 2 briefly explains the industry's technology, cost structures, historical organization in Canada and the United States, cycles of prosperity and distress, and ability to modernize. An understanding of steelmaking technology is necessary if we are to understand as well the industry's crucially important cost structure. Explaining that technology will also show, to the benefit of concreteness, what the industry actually does. That in turn will enable us to understand, in practical terms, matters that can easily become abstract and hypothetical (and often appear so in the formal literature).

Pricing and market capturing strategies are the topic of Chapter 3. Behaviour in oligopolies scarcely resembles standard economic models of competitive markets. It is inherently strategic, and the stakes are ultimately zero-sum: even in expanding markets, what one producer gains another loses. Discounting prices is one strategy for capturing share in export markets, and this may involve price discrimination – selling the same good at different prices at home and abroad. That raises the questions of when state intervention is justified and whether particular industries are especially subject to injury. Economists have devoted much attention to both matters. Demands for protection have interested political economists, whose perspectives conclude the chapter.

A standardized and portable technology has enabled the steel industry to migrate from its original centres in Europe and North America and become a significant exporter. It has also empowered steelmakers to challenge incumbent producers in their home markets. Chapter 4 summarizes current international patterns of steel production and use, explains economic geography's account of industrial location and the basis of trade, outlines the development of steel production and exporting in Japan, South Korea, Brazil, Russia, and China, and considers respective cost advantages. To place these matters in a more general perspective, the chapter concludes with international political economy's views of technological diffusion, the relationship between state and industry, and the international division of labour.

Global steel gluts are a real prospect. Given the scale and speed of expansion, currently in China and prospectively in India, and given the reality that markets rise and fall, the potential volumes in store would test even the most robust producers' ability to withstand surpluses and low prices. The use of trade to unload overstocks combines badly with the steel industry's tendency toward unstable finances; the result is a contingent and worrisome milieu. Chapter 5 discusses the current pattern of exportable surpluses and considers three survival strategies: protectionism, consolidation, and forming close relationships with major customers.

The use of anti-dumping laws to garner the protections once available through tariffs has been criticized by economists. Cases, they argue, are too easy to win. For industries facing threatening import levels, however, that makes anti-dumping actions an attractive expedient. A review of a sophisticated economic literature evaluates the practice and its current suitability for the steel industry. Global consolidation is well underway and holds out the prospect of coordinating prices with outputs. Many in the industry expected that Mittal's merger with Arcelor would start a bandwagon. Whether other majors such as Nippon Steel and ThyssenKrupp will conclude blockbuster mergers like Mittal's remains to be seen. So also is whether consolidation will actually deliver co-operative outcomes. One limit is conflicting self-interest among producers; another limit is competition law. A third strategy is to exploit technological advantages at the top of the product line, where the largest customer is the world's automobile industry and the demanded product is specialty steel. In a world market of standardized commodity steel products, volatile prices, and competitors joining the lower end of the market, this represents the industry's most promising opportunity. An assessment of that prospect concludes the chapter.

Chapter 6, employing differing assumptions about costs of trade, presents two alternative views of the steel industry's global future. In the first, the present difference between international and domestic trading costs remains, with international trade being more expensive. In the second, the difference disappears, with international trade becoming as cheap as domestic trade. Sharply divergent outcomes emerge. With trading costs unchanged, the industry's current level of international activity simply continues. An indicator of how globalized the industry is even now is that one-third of world steel production is exported. The important implication, however, is that the industry's current approach to organizing world production may not change significantly. In the second outcome, globalization progresses to a seamless world in which, because of fallen trade costs, specialized facilities serve international markets. Inherent in both outcomes are the standardization and utility that make steel a naturally globalized commodity.

Will the steel industry's present hierarchy of commodity and specialty steel producers remain? Commodity-steel production is now the mainstay of producers in industrializing countries, while specialty steel is a sector in which producers in Europe, Japan, and North America have regrouped after shedding much of their lower-end capacity. For commodity steelmakers, going upmarket is an attractive option, and those serving large domestic economies are positioned to develop the necessary scale, returns, and expertise. A related matter is ownership. Large investment funds with direct or indirect ties to governments have begun entering world financial markets. That represents a return of the state and raises the question of purposes and effects. For the investing state, do purposes extend beyond normal asset management? And for the receiving state, do effects extend to national economic and security interests? A very recent and unexpected complication is higher energy and transportation costs. If those costs were to become permanent, they would impose significant trade disadvantages on remote suppliers. The chapter concludes with four brief forecasts about the future of world steel and five broader implications, drawn from the steel industry's experience, about the world economy.

An eclectic approach has been necessary. Understanding steel's cost structure and operating constraints – a key source of the industry's often fragile finances – requires visits to microeconomics and industrial engineering. Pricing behaviour and market-capturing strategies take one into the precincts of industrial organization. Steel's expansion to new centres in Asia and Latin America becomes clear in light of economic geography's emphasis on economies of scale and transportation and of the state's ability to lock in factors that generate trading advantages. Anti-dumping rules originated in competition law, whose provisions also set potential limits on international consolidation. Insights from organizational interdependence clarify the industry's prospects of forming close ties with major customers.

An advantage of an eclectic approach is that it brings together literatures whose specialized focuses and methodologies more often keep them in separate realms and, for non-specialists, constitute barriers to entry. By pooling their contributions, we will be able to treat the industry with the necessary scope, ranging from the cost economics of individual producers to the patterns of global production and trade. Doing so will unify an important but disparate set of developments and connect them to a common industrial base. There have been some general treatments of the steel industry, but they were written in the 1970s and 1980s when it was in crisis and very different from today.[24] The present effort is set specifically in the current context of globalization and international consolidation. In presenting these matters, it aims for the kind of straightforward utility that characterizes steel itself.

2
A Tough Industry

The steel industry's technology and economics account for its cyclical nature, its vulnerability to price swings, its structure of large firms, its ability to migrate to new centres of production, and the ability of industrializing countries to become steel exporters. This chapter describes the industry's technology, sectoral organization, products, cost structure, and competitive environment. Because all of these matters bear directly on pricing, trade exposure, and viability, they are important to understand. To the advantage of interest and comprehension, they are also strikingly plain and practical. This chapter, which shows their pivotal importance, is intended as an overview for generalist readers.

Material inputs and technology divide the steel industry into two sectors. The first is iron-ore–based, with blast furnaces smelting the ore into iron, which is then refined into steel and rolled into final products. The second is scrap steel–based, with electric furnaces melting the scrap, which is then rolled into final products. In effect, the distinction is between producing new steel, which is what iron-ore–based mills do, and transforming existing steel, which is what scrap-based mills do.[1] The iron-ore–based producers are in the integrated sector of the industry, whose large installations have long dominated industrial landscapes and earned the sobriquet of "big steel." The scrap-based producers are in the minimill sector, and the small scale of their operations, together with their dispersal in regional markets, makes them much less visible and familiar.

That basic division accounts for important differences in costs and market behaviour as well as for the two sectors' respective successes and difficulties. Cost-based competition has affected integrated producers in all of the world's mature industrial economies, and the competitors have been foreign producers and domestic minimills. Because the integrated sector has been at the centre of trade complaints, has suffered the most from oversupply and falling prices, and has a cost structure that makes for unsteady finances, it is the focus of this study. Minimills are included briefly and comparatively to show the effects of a simpler and cheaper technology.

Making Steel

Raw Iron

Steel is refined and purified iron, and the heart of integrated steelmaking is the blast furnace. It produces iron by combining processed ore with coked coal and limestone, with the coke serving as fuel and the limestone as flux. All iron ores are oxides, and the smelting process removes the iron oxides and other impurities, chiefly silica, aluminum, phosphorus, and sulfur, leaving molten iron. A blast furnace is a tall vertical vessel in which the ingredients are introduced at the top and the molten iron and flux are drawn off at the bottom. The height of a modern blast furnace is 30 metres and the inner diameter is 14 metres. Heated air, generated from recovered exhaust gases, is blown into the base of the furnace, supplying heat and oxygen for combustion of the coke and for a chemical reaction that forms carbon monoxide to remove the oxides from the ore. The limestone combines with silica and other acid-based impurities. The ingredients descend in the furnace as combustion proceeds, becoming increasingly soft as they break down under rising temperatures, which reach 1600°C at the base of the furnace. There the limestone and its absorbed impurities separate into slag, leaving molten iron. The molten iron can be cast into blocks, called pigs, or transferred directly to an adjoining steel mill. Fresh charges of coke, iron ore, and limestone are added at the top of the blast furnace to replace the removed slag and molten iron.[2]

Blast furnace capacity has increased over the past century, with current units, which have 2,000 cubic-metre capacities, producing some twenty times the output of those in 1900. In a modern steel mill, producing iron for transformation into steel is by far the largest cost category, accounting for 55 percent of the total cost per tonne. Adding scrap metal, alloys, and oxygen (whose uses will be seen momentarily) brings the cost to fully 82 percent of the total. In comparison, the cost of labour constitutes only 4 percent, partly because of productivity gains through automation.[3] Because iron making is a continuous process, blast furnaces are expensive to shut down and restart. They do, though, require periodic replacement of their firebrick linings; thus, steel complexes often have two furnaces in order to maintain uninterrupted production. Relining a large, modern blast furnace can cost up to $100 million.

An alternative process known as direct-reduction iron making (DRI) uses gas instead of coke as fuel. Because the process works at temperatures below iron's melting point, it requires higher-grade ore, and its product, sponge iron, requires further processing to remove slag. Compared to blast-furnace

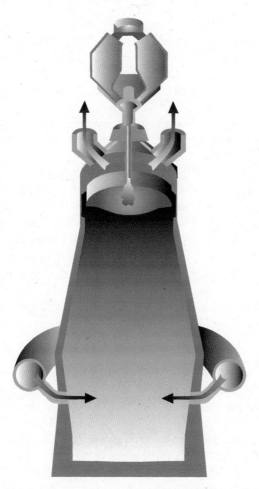

Figure 2.1 Blast furnace. Iron is refined in a continuous process as the ingredients descend in the furnace under increasing temperatures. *Reproduced with permission of Corus Steel.*

technology, DRI requires a lower capital outlay and is efficient at smaller volumes. Its limitation is gas prices. Some minimills in Canada and the United States installed DRI to reduce their dependence on scrap; then high gas prices "devastated their direct-reduction plants, and there are currently no gas-based DRI operations in these two countries." Gas-price differentials actually prompted one DRI plant to be moved from Louisiana to Trinidad. At the same time, cheap gas has made the Middle East an "El Dorado for DRI." Indeed, the world's largest DRI facility today is in Saudi Arabia. Between 1970 and 2004, world DRI output grew from 790,000 to 54.6 million tonnes.[4] To put that figure in perspective, world steel production in 2007 was 1.34 billion tonnes.

A new technology known as smelting reduction substitutes coal for coke and reduces energy and capital costs by eliminating the expensive coke-making process and its environmental compliances. One smelting reduction technology, the Corex process, adds a second stage that removes residual impurities from the sponge iron produced by the first stage, yielding hot metal of the same temperature and composition as that from a blast furnace.[5] An even newer process, Ironmaking Technology Mark 3, uses a mixture of pulverized coal and finely ground iron ore to produce pellets, which are fed into a rotary-hearth furnace and then through various temperature stages, producing nuggets of 97 percent iron. That high purity allows them to be used as substitutes in steelmaking for pig iron and premium-grade scrap. The technology was developed by Japan's Kobe Steel and is being installed in a pilot project in Minnesota. A key investor is Cliffs Natural Resources (previously known as Cleveland Cliffs, Inc.), the major supplier of processed iron ore to the American and Canadian integrated producers.[6]

If new technology ever produces efficient alternatives to blast furnaces for making pig iron in large volume, then the minimill sector, taking advantage of its low capital costs and its ability to serve local markets, and escaping the limits set by scrap-steel supplies, will likely expand accordingly. Required for this would be an additional stage to transform the iron into steel, but that technology is in the early stages of application, as will be seen shortly. Were it to be widely adopted by minimills, the implications for integrated producers could be sombre.[7] At the same time, the cost-efficiency of blast furnaces has been improved by replacing some of the energy provided by coke with that provided by injected coal and oil. Mini-blast furnaces, which cost only $17 to $19 million, compared to $400 to $900 million for a high-volume unit, are an option in industrializing countries.[8] Because of the outlays required to install the new, alternative iron-making processes, the prospect of further efficiency gains as blast-furnace technology is developed further, its ability to produce in very large volumes, and the impressive economies of scale of the most modern units, other forecasts expect only a small decline, with integrated facilities continuing in operation or even being upgraded and producing most of the world's steel.[9] That is the current pattern, with integrated mills accounting for two-thirds of global production.[10]

China is a case in point. The astonishingly fast development of its steel industry – 24 percent expansion in 2004 alone – has been "powered by the blast furnace."[11] In 2007, 90 percent of China's steel was produced in blast-furnace–supplied mills, compared to 66.3 percent worldwide.[12] The same technology is central to large new projects elsewhere. One is in South Korea, with Hyundai Motors installing a $6 billion blast furnace and mill to escape dependence on steel suppliers. The second project is an $8 billion joint venture in Brazil between China's Baosteel and Brazil's Vale, the world's largest iron-ore producer, to produce semi-finished steel slabs for China.[13]

Baosteel decided against the project in 2005 because of currency exchange rates but resumed it two years later in another Brazilian location.[14] The technology itself is attractive in Brazil. Gerdau Steel SA, Brazil's largest minimill-based steelmaker, entered the integrated sector by acquiring Acominas SA, installing a Chinese-supplied blast furnace and developing two new iron mines.[15] Chapter 4 further discusses Brazil's steel industry and its prospects.

It is now possible to upgrade blast furnaces during periodic relinings and run them for increasingly long intervals. That improvement reduces the need for new blast furnaces both in older facilities and in newer ones in industrializing countries, raising the prospect that blast-furnace–based production will prevail in the short and medium term.[16] On the positive side as well, some European and East Asian blast-furnace–supplied mills are highly cost-efficient, a tribute to their scale economies and process efficiency.[17] The converse, however, is also true: given the enormous cost of building new blast furnaces and given the fact that smelting-reduced iron can be used in both integrated facilities and minimills, other forecasts see smelting reduction displacing blast furnaces over the longer run, particularly in smaller integrated installations, as older and less efficient units are retired.[18] The capital-cost barrier of blast furnaces remains formidable, and the operating scale may be well beyond the capacities of all but very large producers. Referring to doubtful prospects for installing blast furnaces in Southeast Asian mills to reduce their dependence on imported raw steel, one steel manager observed that "these days, if you're going to build a blast furnace, it needs to have a capacity of at least 5 million tons per year or else the project will make no sense. A project of this magnitude is just too costly for most companies."[19] For comparison, Dofasco's production in 2005 was 4.2 million tonnes. In addition, financing on that scale requires a stable price environment, and world steel prices over the past decade have been sharply cyclical. Capital costs and plant scale do indeed figure prominently in steel, as will be seen shortly.

From Iron to Steel

Making steel requires further removal of impurities – mainly carbon, silica, sulphur, and phosphorus – from pig or sponge iron. Pig iron's carbon content of 4 to 5 percent would produce excessively brittle steel, and the amount of carbon removed is determined by the desired hardness. For tool and other hard steels, carbon is reduced to between 1.5 and 0.45 percent, and for the most malleable steels to between 0.25 and 0 percent.[20] Mild steel, which is used for a range of products, including automotive body metal and structural steel, has a carbon content of 0.25 percent. Other metals, such as manganese, chromium, and nickel, are added as alloys to produce specific qualities. Before modern blast furnaces were invented, iron was made in small charcoal and coal furnaces. The first steel was made by heating iron with charcoal in sealed crucibles, and its properties were described in 384 BC by Aristotle.[21] In more modern times, production was a small-volume "craft" process requiring pig iron to be finished in small batches, first into wrought iron and then into steel. That made the metal scarce and costly until mass production became possible with the invention of the Bessemer converter in 1855, which reduced the cost of producing steel by between 80 and 90 percent.[22]

Bessemer converters were large, open-ended vessels. Molten iron was poured in, and conversion occurred when air was blown into the batch through openings at the bottom of the vessel. The resulting combustion burned off first silicon and then carbon in a rapid and spectacular sequence. The advantage of the Bessemer process was its speed – twelve minutes per batch. Its disadvantage was its limited ability to remove phosphorus, which was present in high concentrations in North American ores and required expensive pre-processing, and its relatively small volume of fifteen tonnes. The Siemens-Martin, or open-hearth, process, invented in 1866, removes impurities by blowing flame and superheated air in an alternating sequence across a large, shallow pool of molten iron.[23] The open hearth's advantages over the Bessemer process were a forty-times larger volume, which formed the basis for the very large steel complexes that began to be built at the turn of the last century, and its ability to refine iron with a higher phosphorus content. An open-hearth cycle, known as a "heat," lasted from seven to nine hours, but compared to the Bessemer process, high volume offset that cost. The Linz-Donawitz or basic oxygen furnace (BOF), invented in Austria in 1948, uses a large, covered vessel and refines the molten iron by blowing in pure oxygen through an inserted lance. Prompting that invention was a postwar shortage of coal for fuelling open hearths. The BOF actually represented a full application of the Bessemer process' chemistry, which had been limited until the 1940s by the expense of pure oxygen.[24]

To remove impurities further, limestone and other fluxes are added in both open-hearth and BOF processes and drawn off as slag. Because particular qualities are required for various steel products, it is at this stage that the desired carbon content is achieved and that particular alloys are added. At this stage also, steelmaking becomes batch production, with carbon and alloy content being set according to current mill requirements, which in turn are set by the demand for particular finished steels. Completed batches of molten steel are cast and transferred to their respective rolling lines. Both the open-hearth and BOF processes can combine scrap iron and steel with the molten iron, making scrap a valuable industrial commodity. BOF furnaces can take up to 30 percent scrap, and open hearths can take up to 90 percent. Bessemer converters were limited to 15 percent.

Because of the much greater volumes possible, the open hearth began replacing the Bessemer process at the beginning of the last century and remained the industry's mainstay until producers began converting to BOF

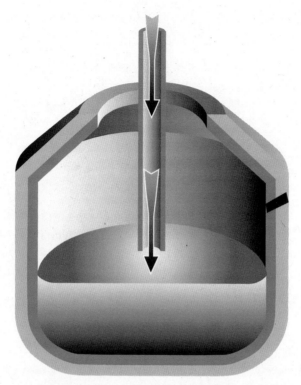

Figure 2.2 Blast oxygen furnace. Iron is refined into steel from a chemical reaction with oxygen, which is introduced through a lance, as shown. *Reproduced with permission of Corus Steel.*

process in the 1950s. BOFs are more efficient than open hearths at all levels of production, and the efficiency gains are the same for both labour and capital. The threshold of scale economies is also relatively low.[25] The source of the BOF's economy is a heat time of only 45 minutes – one-tenth that of an open-hearth furnace – which provides much greater energy and labour efficiency. The improvement has been dramatic: "Melt-shop productivity with BOFs in 1999 is 1,000 times greater than with open hearths in 1920 ... 0.003 manhours per ton vs. 3.143 manhours per ton in 1920."[26] Fuel costs for a BOF are one-sixth those for an open hearth.[27] During the mid-1950s, when the new technology was first being implemented, a 500,000-tonne BOF mill reduced total costs per tonne by $6.46 and variable costs by $4.56. That translated into a 15.2 percent cost and price advantage.[28] These gains made adopting BOF the steel industry's single most important improvement of the last century.[29]

The world's first steelmaker outside Austria to adopt BOF for commercial production was Canada's Dofasco in 1954. The same year, McLouth Steel became the first American adopter. Japanese steelmakers made the first large-scale installations as they expanded in the 1950s. Because the technology was new and unproven in high-volume applications, American and European steelmakers were much slower in adopting it. The Japanese were attracted by the process's large savings over open-hearth technology, an important consideration in light of their more limited investment resources and their interest in low production costs. Another reason was BOF's lower use of scrap, which was scarce in Japan. American and European steelmakers did not begin adopting BOF until the mid-1960s, giving the Japanese steelmakers a strong initial cost advantage as they began entering world markets.[30] As will be seen later, one of the investment decisions facing steelmakers was whether to incur the significant capital costs of this new technology. By 2007, open hearths represented only 2.5 percent of world steel production. The only surviving installations are in Ukraine, where they still produce 44.8 percent of output; in Russia, where they produce 16 percent; in the other Commonwealth of Independent States (CIS), where they produce 8 percent; and in India, where they produce 1.9 percent.[31] Using their sizeable export profits, Russian mills have been retiring their open hearths.

Electric Furnaces and Minimills

The electric furnace, patented in 1878 by William Siemens, produces steel by melting steel scrap and removing impurities. Although designs differ, the basic unit is a large covered vessel with openings on the side for introducing scrap and for drawing off slag and molten steel. Electrodes, extending into the vessel from the cover, melt the scrap with heat created by electric arcs between the electrodes and the steel.

As in open-hearth and BOF furnaces, limestone and other fluxes are added to electric-furnace batches to remove impurities and are drawn off as slag. When scrap is being processed that contains "tramp" metals such as copper, chromium, and nickel, directly reduced, smelting-reduced, or commercial pig iron is used to dilute them. Removal of those metals would require expensive pre-processing.[32] The degree of dilution possible without compromising the quality of the produced steel sets the limit for iron that can be substituted for scrap.

Until 1940 electric furnaces were limited to producing high-quality specialty steels. Their use expanded rapidly when it was discovered that they could supply small mills with steel at prices below those of larger producers that used blast furnaces and open-hearth or BOF facilities. Electric furnaces became the core of a new steelmaking method and the basis of minimills. Because minimills use steel that has already been produced, they are spared the integrated mills' very large cost of refining iron ore into steel. One important saving is energy. An electric furnace processing only steel scrap – with no pig iron or other substitutes added – consumes just 10 percent of the

Figure 2.3 Electric arc furnace. Scrap steel and other ingredients are melted by the heat from an electric arc generated between the two vertical electrodes. *Reproduced with permission of Corus Steel.*

energy of a blast-furnace/BOF process.[33] Electric-furnace technology has progressed accordingly. Heat times in 1940 were six hours; now they are under an hour. Volume has expanded, with current units producing up to 150 tonnes per hour.[34]

Although a major cost advantage depends on the price of electricity, technological development is finding ways to reduce electricity use by pre-heating the metal charge with reused hot gases. A recent innovation is a two-shell technology that uses a second BOF vessel to burn off carbon and other impurities with an oxygen lance. Oxygen can also be injected directly into electric furnaces, but the purpose of this is more to raise temperature than to remove impurities. If the two-shell process becomes feasible for widespread application, minimills will be able to refine pig iron and other scrap substitutes into steel, reducing their dependence on scrap and dupli-cating what the integrated mills do.[35] Compared to the other methods, electric furnaces can be operated at more steady temperatures, making it possible to add purifying slags and produce highly refined steels. A ladle furnace is used to homogenize the temperature of steel from an electric furnace and remove additional impurities. A subsequent process, vacuum degassing, adds further refinement and ductility.[36] For electric furnaces, the limit on purity is the presence of tramp metals in steel scrap. Electric furnaces also have moderate volumes compared to BOFs, although that does make them suitable for smaller local and regional markets.

Forming Hot Metal

Integrated mills combine all the stages of steelmaking at a single location. Blast furnaces transform the raw ingredients into iron, which supplies BOF furnaces. Molten steel is cast into large intermediate forms – blooms, billets, and slabs – which in turn supply rolling mills, which transform them into final steel products.[37] A basic distinction can be made between "long" and "flat" products. Long products are rolled from blooms and billets. A bloom has a rectangular cross-section of 16 centimetres or more and is rolled into structural beams; a billet has a square cross-section of 4 to 14 centimetres and is rolled into bars, rods, and wire. A slab has a flat cross-section and is rolled into steel plate and coiled sheet steel. Pipes are made from welded steel plate, and seamless pipes and tubes are made by a process that pierces hot billets. For convenient transport, coiled sheet is produced in large rolls, hence the term. Coil is one of the most widely used steel products, and two of the largest customers are the automobile and appliance industries. Yet another distinction is between hot- and cold-rolled products. Hot rolling produces a rough surface; cold rolling at ambient temperatures adds strength to the metal and makes its surface smooth and lustrous. Cold rolling also produces more exact and consistent tolerances. Rolling mills shape these products in a succession of stages as the steel passes through smaller and smaller clearances in a long line of rolling stands. The first hot strip mill was installed in 1924 and the first cold-rolling mill in 1939.

Long structural products can be made to relatively low standards because their primary requirement is strength. Compared to flat products, they can also be produced cheaply; thus they constitute the low-value–added end of the product line. Flat products, particularly specialty sheet steels, are made to much higher standards for demanding customers. The automobile industry, a principal user, fabricates car bodies from sheet steel stampings, and the need for lightness, strength, corrosion resistance, flawless surfaces, fine tolerances, formability, and special coatings has driven product development. Termed exposed-sheet steels because they are used in surface body panels, these products involve sophisticated metallurgy and expensive processing. Along with specialty and alloy steels, they are high in added value. Developing new steels for the world's automobile industry is one of the integrated sector's most promising means of survival, as will be seen in Chapter 5.

In the past there was an intermediate stage in which the molten steel was cast into large ingots, which were transformed in a primary mill into blooms, billets, and slabs. These were then sent on to rolling mills for processing into their final forms. Continuous casting, a technology developed in the 1950s, has eliminated the primary mill's large capital, labour, and energy costs by forming the molten steel into blooms, billets, and slabs directly. In North America's first successful application of this new technology, Atlas Steels of Welland, Ontario, developed a continuous caster for 18-inch strips in 1954.[38]

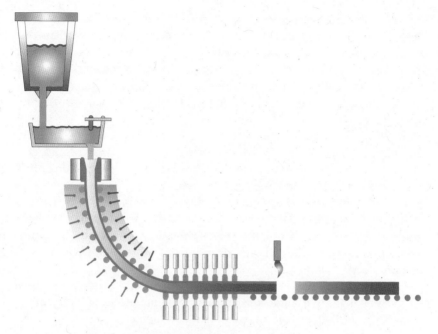

Figure 2.4 Continuous Caster. Molten steel is introduced at the top of the caster, which rolls it to smaller thicknesses as it descends and cools. *Reproduced with permission of Corus Steel.*

Initial cost savings, as the technology was being introduced in the 1970s, were $4.70 per tonne.[39] Continuous casting combines naturally with the short production cycles of BOFs. For that reason, Japanese steelmakers were the first major adopters, having modernized their plants with BOF furnaces. Together, the two technologies lend themselves to continuous-flow production processes. To achieve the elaborate control and coordination needed, the Japanese were also the first to develop computerized production management.[40]

Although American and European steelmakers were initially skeptical of continuous casting's ability to produce consistent metallurgical quality, subsequent development of the technology has yielded results superior to those of the previous casting process. The technology has now been adopted worldwide and in 2007 accounted for 92.4 percent of total production. Again the exceptions are Russia, Ukraine, and the other CIS states, with 71.2 percent of Russia's output being continuously cast (up from 54 percent in 2005), and 34.3 and 99.1 percent respectively for Ukraine and the other CIS states.[41] For improving efficiency, continuous casting has been second only to BOF. Canadian and American steelmakers had converted to continuous casting by the end of the 1980s.

Thin-slab continuous casting was developed commercially in 1983 to improve the production of sheet steel. The process saves intermediate processing costs by reducing slab thickness from 16 centimetres – the size produced by previous technology – to 4 centimetres. That eliminates several stages of hot rolling and reduces the length of finishing lines, allowing energy, labour, and capital savings and smaller facilities.[42] Because of these advantages and because the technology is efficient at small volumes, minimills were quick to adopt it. For them it opened the way to flat-rolled products and represented a major step upmarket, although at the high end, metal quality is a limitation. High-quality cold-rolled and stainless sheet requires steel free of tramp metals. In automaking, for example, sheet steel from a minimill would be of suitable quality for a seat back but not for a roof panel. "Severe metallurgical shortcomings" are the barrier.[43] For the integrated steelmakers a limitation of continuous casting was a large capital cost at a time when their profits were under pressure from minimills and imports.

The goal of current technological development is to integrate the stages of steelmaking into a single and uninterrupted process from blast furnace to finished steel. An important part of that development has been the direct hot-charge process, which feeds freshly cast slabs immediately into sheet rolling mills, saving reheating costs.[44] The current stage of development is a sheet process that eliminates slabs altogether and that casts molten steel directly into strips 1 to 4 millimetres thick. Because steel cast so thinly solidifies quickly, the engineering challenge is to control the distribution of the molten metal and keep it from escaping at the edges of the mould. Prospective cost reductions are significant: sheet casting would eliminate most of the rolling process and could represent a three-quarters reduction in the costs of capital equipment.[45]

Beginning with the introduction of computerized controls in 1963, steelmaking has become increasingly automated. In the words of a vice-president of SMS Schloeman-Siemag AG, the developer of thin-strip continuous casting, technology is moving toward full automation and continuous flow. Steel production, he predicts, "will become more and more ... like the chemical industry, where it's going automatically. We are coming to that step by step ... to the point where human beings are only watching the process and thinking about how to make it better."[46] Ever since the technological breakthroughs of BOFs and continuous casting, the steel industry's innovations have been in production control, that is, in developing ways (a) to quickly adjust individual processes to maintain exact specifications and (b) to shift output among various product lines.[47] Computers are expected to take over production management. One forecast expects computerized control of "much of the steel business ... except for the price setting. The computer reviews new orders from the company's customers ([transmitted] by computer, of course), determines the steel production and rolling schedules, controls the processing of the product and keeps the customer informed as to the status of each order."[48]

Automation's cost savings have been impressive. Between 1982, when the new technologies were becoming widely adopted, and 1995, worker productivity more than doubled from 10.1 man-hours per tonne of finished steel to 3.9 man-hours. Some facilities are able to produce a tonne in one man-hour. As related advantages, synchronized continuous production allows mills to reduce inventories, eliminate lead times, and respond more quickly to customer orders. As investment in new technology continues, so do improvements in productivity. In the Canadian steel industry, productivity per worker increased by 74.4 percent between 1990 and 1997.[49] In the United States, restructuring in the steel industry during the same period increased productivity at a faster rate than in the manufacturing sector overall.[50] As one index, the price of American-produced steel fell by 30 percent between 1984 and 1997.[51]

Integrated steel producers operate at single sites that receive coal, iron ore, and limestone in large bulk and ship out full arrays of finished steel. These are enormous installations. A plant with 5 million tonnes of annual capacity – a moderate volume by current standards – will require 6 million tonnes of iron ore, 3 million tonnes of coking coal, and 0.9 million tonnes of limestone per year. Deliveries of materials and shipments of finished goods will total more than 13 million tonnes.[52] Facilities of this scale represent large and concentrated capital investments, a fact that has significant implications, as will be seen shortly.

Costly Ingredients
In North America, integrated mills were built in the late nineteenth and early twentieth centuries around the Great Lakes, where it was cheapest to combine heavy volumes of coal from Pennsylvania and the central Appalachians with iron ore from Minnesota and the Upper Peninsula of Michigan. After the Second World War, North American steelmakers developed iron mining in Labrador. The Wabush Mine in Labrador City is jointly owned by Cliffs Natural Resources, ArcelorMittal Dofasco, and United States Steel Canada (formerly Stelco). Cliffs Natural Resources, the largest supplier of processed iron ore to American producers, also owns mines in Minnesota and northern Michigan. Rio Tinto Group, the world's second-largest iron-ore producer, in 2000 acquired part ownership of the Iron Ore Company of Canada, which produces in Labrador. The Quebec Cartier Mining Company came under Dofasco's ownership in 2005, the year before the latter was purchased by Arcelor. In August 2007, United States Steel Canada and Cliffs

Natural Resources agreed to sell their stakes in Wabush to ArcelorMittal Dofasco. The two withdrew their offer in March 2008 and in October announced the termination of discussion of the matter with ArcelorMittal Dofasco. Algoma Steel had purchased its iron ore on contract from Cliffs Natural Resources. When Essar Steel acquired Algoma in 2007 it also bought Minnesota Steel, which owns iron-ore–mining facilities in the Mesabi Range and is constructing an on-site plant to produce iron pellets. Essar plans to use the short shipping distance to supply Algoma.[53] Interestingly, Essar is also constructing a steel mill on the Minnesota ore site that will use gas-fired DRI to supply an electric furnace. Given the role that gas prices played in eliminating DRI in North America, the offsetting cost advantage for the facility – billed as the most cost-efficient on the continent – is a proprietary on-site ore supply. From a world market perspective, these ore sources supply regional steel industries and free North American mills from dependence on overseas sources. That fact has made them attractive to foreign buyers, as shown by Arcelor's, Essar's, and Rio Tinto's acquisitions.

In contrast, steelmakers in Europe and Asia must import iron ore, and the world's highest grades are mined in Brazil and Australia. Because processing costs make the value of ores dependent on their iron content, rich ores can bear long shipping distances. The feasibility of exporting them depends on efficient extraction and transportation to ocean ports.[54] Both Brazilian and Australian ore producers enjoy these advantages. Inexpensive ocean bulk shipping has reduced delivery prices and created a world market, although energy prices, which rose steeply in 2007, may seriously affect transportation costs if they return to these high levels from their collapse in the fall of 2008 (see Chapter 6). Mills have been able to depend heavily or entirely on distant sources, and the newest integrated facilities, chiefly in Europe, Japan, South Korea, and China, have been built at seaports.[55] Of the world's major producers, ArcelorMittal depends on imports for about half its ore supply, partly because the firm's acquisitions have included sources of iron ore. In contrast, China's Baosteel is 80 percent dependent on imports and South Korea's POSCO is 90 percent dependent.[56] The EU relies on imported iron ore for 83 percent of its supply.[57] In the early 1980s, Brazilian ores were offered in the Great Lakes region for lower delivered prices than local ores, forcing the North American producers to cut costs quickly. They did so by sharply improving productivity, and they have remained the Canadian and American steelmakers' primary source because of competitive prices and convenient supply.[58] This provides an important advantage over import-dependent producers.

Consolidation in the mining industry has left three global producers – Brazil's Vale and the British/Australian firms Rio Tinto Group and BHP Billiton PLC – and together they control about 75 percent of world supply.

They sell to steelmakers on the basis of annually negotiated contracts, and the price struck with Japan's steelmakers has informally set the pattern for the others. The three exercise very strong market power. In 2005, they were able to extract a 71.5 percent price increase, the steepest in twenty-three years. Then in 2006 they extracted another 20 percent, and in 2007 an additional 9.5 percent. Ore supply and demand set the parameters. By 2005, China's need for imported iron ore for its burgeoning steel industry had outstripped projections and squeezed world supply, allowing the producers to negotiate steep price hikes. These price increases raise costs for all buyers, and for those who are heavily import-dependent the costs are impressive. For 'China they amount to some 2 percent of GDP.[59] To reduce their exposure, Baosteel and other Asian producers have entered joint development projects with Brazilian and Australian mining and resource firms, but direct purchases continue to be significant. In 2008, the Chinese Metallurgical Mining Enterprise Association said that China's ore imports might rise in the coming year by 14 percent to 435 million tonnes.[60]

In February 2008, Vale negotiated a 65 percent price increase with Japan's Nippon Steel and South Korea's POSCO. Rio Tinto was not willing to treat that price as the industry benchmark, and in June it announced that it had obtained a 96.5 percent increase with Baosteel, which was negotiating on behalf of all of China's steel producers. The rationale was increased ocean transportation costs. For ore delivered in China, these costs represent about 30 percent for Australian ore and, because of a much longer transit, nearly 50 percent for Brazilian ore. Rio Tinto argued that the difference amounted to a bonus for Australian ore, which the new price should reflect. The informal benchmark price arrangement may itself be in danger of being undercut, with Rio Tinto shifting sales away from yearly contracts to the spot market to take advantage of rising prices.[61] That advantage shifted rapidly, however. Collapsing demand for steel in the fall of 2008, together with transportation costs that had declined because of falling energy prices and demand for ship capacity, caused iron ore prices to fall by 50 percent, and the spot market prices were actually below those contained in existing contracts.[62] Nonetheless, Rio Tinto's success in getting a nearly 30 percent premium over Vale's contract price shows the power available among the current three producers. To the benefit of North American steelmakers, China's dependence on imported ore, together with its transoceanic costs of shipping steel, offset its labour-cost advantages.[63] More will be seen about respective labour costs in Chapter 4.

Being independent of the iron-ore cartel is indeed important.[64] Sharing that benefit with the United States and Canada are Russia, Ukraine, and Brazil. Russian steelmakers, who have consolidated former state-owned producers into four large firms – Evraz, OAO Severstal, Magnitogorsk, and

Novolipetsk – have used some of their export earnings to buy iron-ore and coal producers in Russia and Kazakhstan. The Russian steelmakers have labour costs among the world's lowest and owning their supplies of iron ore and coal enhances that advantage (e.g., over China). All four producers regard their owned supplies of ore and energy as assets in prospective mergers with steelmakers in Europe, Asia, and North America; at the same time, ore supplies make them attractive acquisitions.[65] Access to mining assets was a central consideration in Mittal's purchase of a 93 percent stake in Ukraine's Krivorozhstal in 2005, a deal also pursued by Arcelor.[66] The following spring, Arcelor, seeking to thwart Mittal's hostile takeover bid, put forth a plan to merge with Russia's Severstal. The union would have joined Arcelor's sophisticated expertise and technology with Severstal's supply of iron ore. In 2008, Corus Steel, which had acquired its ore in the open market, announced plans to open mines in Canada and South Africa.

Coking-grade coal is mined in quantity in the central Appalachians in the United States (the source of the world's premium grade), in Alberta and British Columbia in Canada, and in Queensland in Australia. All are suppliers to the world steel industry. Russia and Ukraine mine their own coking coal. Germany's coal industry is shrinking rapidly, and with the government's announcement in 2008 that it would end coal subsidies, the industry is expected to disappear over the next ten years and leave steel producers completely reliant on imports. Ocean transportation makes coal available globally, and as with iron ore, prices are set in negotiations. The price struck between Japanese steel producers and Australian coal suppliers sets the pattern for the rest of the industry. Linking ore and coal prices is a negotiating option, and in 2006 the Japanese producers succeeded in getting an 8 percent coal price reduction to compensate for the 20 percent hike they were forced to accept from their ore suppliers.[67] Coal prices have also been rising, reaching a peak of $300 per tonne in 2008 before falling to less than half that figure by the end of the year with the decline of steel demand from the global economic crisis. Volatile coal prices have interested steelmakers in purchasing coal producers, and ones in North America are regarded as prime candidates. ArcelorMittal in particular is "moving aggressively to boost its self-sufficiency" and in 2008 purchased three coal mines and processing facilities in Russia for $720 million.[68]

On the output side, an advantage held by the Canadian and American steelmakers around the Great Lakes is that they are close enough to their largest customers, the automakers, to be regarded as local suppliers that can provide quick deliveries and direct communication.[69] The automakers prefer to deal with local steel producers because they can coordinate production schedules, collaborate in product development, and operate closely linked supply chains. In North America, that preference benefited American and Canadian steelmakers as imports penetrated other segments of their markets.

Industrial regionalization, supply linkages, and world trade will be discussed further in Chapter 4.

Minimills: Smaller, Simpler, and Cheaper

Minimills operate on a smaller scale because their source of steel is electric furnaces, which have modest capacities compared to blast furnaces. Electric furnaces are economical and efficient at relatively small volumes, as are continuous casters. Minimill plants consist of an electric furnace, a continuous caster, and a rolling mill. Small facilities mean low capital costs, and minimills have attracted ample financing. Because they entered the industry when the integrated sector was mature, they were able to adopt the latest casting and rolling-mill technology from the outset. Although the minimills "have rarely been technology leaders ... they have taken full advantage of existing technology to exploit market opportunities. Their success is based on choosing the right technology for the right market [and] organizing the workplace to take advantage of that technology."[70] A simpler steelmaking process has allowed minimills to sidestep the integrated mills' long history of rigid job classifications. That has allowed more co-operative working arrangements and has also reduced coordinating costs, although the integrated mills closed some of that margin as they reduced their workforces and trimmed their lower-end product lines.[71] Moreover, the minimills' flexible production processes enable them to respond quickly to shifting demand.

Because scrap is the material input, minimills do not need molten or cast steel from integrated mills. Moreover, minimills were organizing just as the integrated producers were converting from open-hearth to BOF technology. Because BOFs can take less scrap than open hearths, the integrated producers' demand for scrap fell, lowering its price for minimills and providing them with an important cost advantage.[72] Costs do, however, depend on recovery and processing. Limited supplies of higher-grade scrap – chiefly from steel mills themselves and from manufacturing processes using steel – raise costs because of scarcity and because of the expense and lesser yield from processing lower-grade scrap, such as that from discarded automobiles. That scarcity increases as both steel mills and manufacturers adopt technologies that reduce waste. As well, because scrap is a bulky and heavy commodity, transportation costs affect minimill advantages. Scrap is not abundant worldwide, and countries with shortages must import it. A major source of export scrap is North America. The largest importers in 2007 were Turkey, South Korea, Spain, Taiwan, and Italy, which together represented 44 percent of world imports. The United States, the CIS states, Germany, Japan, the United Kingdom, the Netherlands, and Canada were the largest exporters, representing 72 percent of the world total.[73]

Scrap may be supplemented with directly reduced or smelting-reduced iron, as was seen earlier.[74] That places minimills in the position of either producing a key input themselves or buying it. To hedge against future shortages of scrap, a few minimills in Canada and the United States added DRI facilities; however, they abandoned them when gas prices rose. Even while DRI was being used, its small scale in North America reflected a history of unusually low scrap prices.[75] Scrap supply, however, has not increased, and demand is growing, particularly in Asia. Scrap prices, responding to conditions in a competitive world market, are quite changeable. Global scrap prices rose from $150 per tonne in 2000 to $270 per tonne in 2006, reducing an important cost advantage.[76] Scrap prices began to surge in 2007, and in June 2008 they were $650 to $660 per tonne before falling to $400 a tonne in fall 2008.[77] Unless minimills adopt the new electric furnace/BOF technology, they will not be able to produce new steel and will have to depend on scrap and raw iron.

The minimills' other exposure is electricity prices, which are often their single largest cost of production. Electricity rates vary by locality and are subject to various influences, including government subsidies, making them an important locational factor. Although new electric-furnace technology is more energy-efficient, higher electricity rates would still reduce an important advantage, and over the longer term that cost, as with scrap prices, is relatively unpredictable. Also, electricity supply from power grids is subject to restrictions during summer heat waves. At the same time, since minimills do not depend on the integrated producers for their raw metal, they can locate near their markets, eliminating the need for expensive overland transportation and encouraging responsiveness to local demand. Minimill managers have indeed been responsive. As leaders of new firms in an emerging sector, they have been flexible and imaginative, quick to grasp market opportunities and adopt new technology.

Overall, the minimills' net advantages have enabled them to capture the predominant share of the structural steel market from the integrated producers. In product lines where minimills compete directly with integrated producers, minimills establish the effective price.[78] Nucor, the largest North American minimill, along with United States Steel and ArcelorMittal, is one of North America's top three producers. Minimills now account for 58.9 percent of total steel production in the United States and 40.8 percent in Canada. Worldwide, 31.2 percent of total steel output is produced in electric furnaces. By percentage of total output, the EU is the largest producer, followed by the United States, China, and Japan. One indicator of China's prodigious output from integrated mills is that electric furnaces account for only 10 percent of its total production. In the industrializing countries, the highest percentages of electric-furnace production relative to total output are in India, Turkey, and Mexico. Electric-furnace technology's ability to substitute for imported coal and to serve moderately sized markets is evident in the countries where its production share is highest – Venezuela and Saudi

Arabia at 100 percent, Egypt at 83.9 percent, and Iran at 77.3 percent. In the industrial world, Spain (at 77.9 percent) and Italy (at 63.3 percent) are the largest users. Greece, Luxembourg, and Portugal produce all of their steel in electric furnaces, but their combined output is just over one-third of Spain's.[79]

In North America, minimills entered at the same time that steel imports were beginning to gain market share. The North American minimills' output grew to some 23 million tonnes between 1974 and 1994 – exactly at a time when the world market was producing excess capacity, both from expanding producers in Asia and Brazil and from obsolete facilities in Europe and North America that had yet to be closed. For American integrated producers, the minimills' gain was their direct loss.[80] Canadian integrated producers were less severely affected. Historically they had concentrated on particular market segments and had left unserved ones for imports. That same market division provided some latitude to accommodate minimills, although their expansion was one reason for the integrated producers' straitened financial condition at the beginning of the last decade, which saw Algoma Steel declare bankruptcy in 1990.

For many products, including steel pipe for the petroleum industry, Canadian minimills serve regional and specialized markets, and the Canadian industry continues to be less clearly divided than the American industry between integrated producers and minimills.[81] Electric furnaces can be installed in integrated facilities to bypass or augment the blast-furnace/BOF steelmaking process, and ArcelorMittal Dofasco has an electric furnace at its integrated facility in Hamilton, Ontario. Because such combinations are possible, the boundary between integrated and minimill operations may eventually become less distinct.[82] As has been seen, a limit of current minimill technology is metallurgical quality. That has allowed both Canadian and American integrated producers, along with their European and Japanese counterparts, to claim the market for higher-end products and specialty steels. At the same time, it has meant conceding share in the lower and middle-grade range of the market they once easily dominated. That occurred as the integrated producers were also experiencing new competition from imports.

Steel service centres are supplied by the mills and act as pre-processors, fabricators, and distributors. They unwind steel coil to make it usable for manufacturing, cut steel into intermediate lengths and shapes, add coatings, and pre-process it into blanks and sub-assemblies. They can specialize in supplying particular industries and may locate themselves close to principal customers. In the automobile industry, service centres have expanded into producing tailored blanks, which are stamped components of car bodies, and applying special coatings and paints. Seeking to align more closely to the automobile industry, some steel producers have undertaken those activities themselves (see Chapter 5).

Market Behaviour: Costs, Constraints, and Contingency

The integrated steel industry has four characteristics that directly affect its behaviour in both domestic and international markets as well as its financial viability. The first characteristic is high capital and fixed costs. The second is economies of scale. Together these result in high entry and exit barriers and directly influence pricing decisions. The third characteristic is high substitutability among producers and low marginal costs. These, combined with high fixed costs and barriers to exit, invite managers to discount prices rather than cut production. Since steelmakers generally face similar cost and market conditions, the price discounts readily become intramural and mutually destructive. The fourth characteristic is steel's status as an intermediate good facing inelastic demand. That makes the industry immediately subject to economic downturns and to production cutbacks in key durable-goods sectors. Episodes of decreased demand feed back directly into pricing decisions, which in turn are shaped by fixed and marginal costs and by barriers to exit. Together these four characteristics operate with powerful combined force. They account heavily for the industry's proneness to financial distress and for the problems and demands to which that gives rise.

High Capital and Fixed Costs

The integrated steel industry, in the words of one financial analyst, "eats capital for breakfast."[83] Each element of the production process, from blast furnace to final rolling mill, represents a large, highly specialized, complex, and durable capital asset. These elements are available from specialized producers and can be assembled as full systems in completely new greenfield installations or added as individual elements to existing facilities. The size of these elements has grown with increases in capacity, and their complexity has grown with automation. The purchase prices of these elements have climbed accordingly, producing "lumpy investment and high start-up costs."[84] The expense of financing these assets is an important part of an integrated mill's fixed cost, as is the expense of maintaining them. Given the durability of the facilities involved, that fixed cost is long-term. Covering it produces a break-even point that normally requires high capacity utilization.[85] That in turn makes integrated producers highly vulnerable to economic recessions and falling demand, and makes it a risky proposition to develop entirely new greenfield installations, even though they have the advantage of assembling the latest technology as fully integrated systems. Capital markets know this.[86] The same vulnerability makes steelmakers especially bad risks during times when falling competitiveness and profitability indicate the need to finance more efficient technology. That double bind compounds the integrated steel industry's difficulties whenever revenues fall, whether from economic recessions or from competition from minimills and imports, and whenever competitors achieve efficiency gains.

Conversely, when the industry is enjoying strong demand and high prices, as it had been between 2004 and mid-2008, it attracts investors. As an example of rapid upside movement, the share price of United States Steel, reflecting surging global demand, rose from $18 in the industry's crisis year of 2000-1 to $60 in 2006, and in June 2008 it reached a high of $196. The authoritative *Value Line Investment Survey*'s positive assessment of United States Steel in April 2008 was based on rising volume, operating rates, spot-market prices, and "selective steel shortages" in the global market.[87] For the integrated steel sector overall, "the long-term view is promising, as the group again looks like a growth sector, albeit a cyclical one, in an ever-more global industry setting."[88] The principal threat envisaged by *Value Line* is a decline in world steel demand, which would divert foreign steel production to the American market and depress prices.[89] Between that *Value Line* assessment and the subsequent one of October 24, 2008, conditions in the industry had indeed changed, with United States Steel's share price diving to a low of $39. *Value Line*'s assessment stressed that the company had strong finances and low debt that would carry it through to a mid-2009 recovery, but that the industry's cyclicality together with the prospect of a significant recession made the stock "not for the faint of heart" – even though the company's long-term outlook is "upbeat."[90] This rapid change of fortune shows how widely variable prices and conditions in the industry can be. Demand, profitability, and the favourable attitude of the financial markets can all change very quickly.

Producers must stay abreast of the industry in technology and productivity. This is not as simple as it might appear. The long-term durability of integrated plants can be a disincentive to modernize if their expensive equipment is not fully depreciated. That was one reason why American integrated steelmakers delayed replacing their open-hearth furnaces – which have life expectancies of sixty to seventy years – when their European and Japanese counterparts were doing so. As well, modernizing parts of existing facilities is expensive if the upstream and downstream parts also have to be upgraded to realize the potential efficiencies. One reason why American steelmakers delayed introducing continuous casting was that the technology would not have much effect on costs if it were simply spliced onto old open-hearth furnaces and rolling mills.[91] These technological constraints produce "significant indivisibilities" in the production process, and together with the high costs of investment, they make steelmakers "economically fragile."[92]

These facts have several important implications. The first is that revenue must cover fixed costs in the long run and variable costs in the short run.[93] The result is a cost pressure to operate at high levels of output regardless of market conditions.[94] The second implication is that these built-in requirements for high production, combined with capital costs, can complicate modernization decisions. New facilities are poor investments if their capital costs are not covered by the added gains in efficiency and by adequate revenues. That places producers at a long-term disadvantage against rivals who *are* able to cover the costs of those improvements.[95] That was the position of European and American steelmakers in the 1960s and 1970s as Japanese steelmakers made heavy investments to increase efficiency and scale. The second-best option, that is, incorporating segments of new technology into existing facilities, may also not be feasible if those segments' contributions to system-wide efficiency are too small to cover their cost. Failure to stay abreast of the industry, however, means diminishing viability.

Minimum Efficient Scale

Covering high fixed costs requires high rates of capacity utilization to spread those costs over a many units of production.[96] Capacity itself has grown with advances in steelmaking technology. That technology was developed by Japanese steelmakers at the peak of their industry's expansion in the 1960s, and the purposes were to supply metal for rising industrial production and to achieve greater economies of scale. Although the largest modern blast furnaces have a capacity of some 9 million tonnes per year (abbreviated to mmt for convenience), the threshold of full efficiency is lower. Minimum efficient scale is the production level at which the cost per unit of output is lowest.[97] The minimum efficient scale of a BOF shop is some 5 mmt annually, and that of a rolling mill is between 4 and 5 mmt, although in some large and new mills the level can be as high as 8 to 10 mmt. The component with the highest minimum efficient scale sets the output level for the other components, and in recent years the output level has been set by blast furnaces.[98] Modern blast furnaces produce a high and continuous output and require a correspondingly large steelmaking facility to consume the steady flow of iron. The same is true of BOF furnaces, whose short heat times produce high and steady volumes that are fed directly to the casters and rolling mills. These immediately connected stages account for steelmaking's high indivisibilities of production.

To cover their costs, integrated mills until recently operated at 75 to 80 percent of their capacity, with 80 percent the standard benchmark.[99] As the industry recovered from its depressed conditions in the late 1990s, it salvaged and consolidated the most efficient facilities and improved productivity. Record profits in the following decade strengthened financial reserves. Together, these improvements have reduced costs sufficiently to enable the most efficient steelmakers to operate at 65 percent capacity and still remain profitable – giving them the option of responding to depressed markets by cutting production to reduce supply and maintain prices and thereby moderate the potential for mutually destructive price wars, which occur during cyclical downturns when firms maintain output and cut prices.[100] That preferable outcome requires the industry's other producers to cut their own output. If they do not, the firms that do cut output will suffer foregone sales *and* lower prices. Chapters 3 and 5 consider the strategic implications.

Variable costs, which rise and fall with the level of production, are normally relatively constant, so that a producer can increase production to cover fixed costs without incurring rising variable costs. That, too, encourages operation at high capacity.[101] The converse is also true: because fixed costs are so large and variable costs are relatively small, mills have a margin in which they can cut prices and still cover their variable costs.[102] That encourages mills to keep producing in the face of falling prices – a fact with centrally important implications, as will be seen presently. The ratio of fixed and variable costs changes when the price of material inputs – iron ore and coal – rises significantly. When that happens, cutting production lowers the burden of variable costs and makes it possible for steelmakers to maintain price levels by reducing output. Although iron ore and coal prices had fallen from their peak levels of early 2008, they were still high enough by fall 2008 to make this a feasible option, and the major producers had made temporary reductions of 15 to 35 percent.[103] If individual producers can maintain profitability at a lower output, that situation is preferable to maintaining production levels and cutting prices. Lowering output across the industry requires coordination – a practice with a long history in the steel industry, as will be seen shortly.

The operating rate necessary to cover total costs depends on prevailing steel prices and can vary widely. During the volatile early 1980s, minimum operating rates in major American mills varied from 34 percent to 125 percent.[104] The former figure reflects unusually high prices and the latter unusually low ones – an additional 25 percent of production over full capacity would have been needed to avoid financial loss. The variation highlights the exposure of steel producers to price fluctuations as well as the continuing burden of fixed costs.

Maintaining an adequate scale of production also affects sales strategies. "Steel producers no longer concentrate on volume alone; rather, companies bid on specific product characteristics at a price that will use plant capacity and earn a reasonable profit."[105] Successful product specialization concentrates resources in their highest-volume applications. In the words of one analysis of steel productivity, "finding a position in the market that allows you to make the most of the technology and people you employ means operating your machinery at high rates of utilization. The importance of this cannot be overstated. When billions of dollars of capital are tied up in plant and equipment, idle machines translate into huge losses."[106]

Because integrated mills must operate continuously and produce steel in large units, it is difficult to adjust volume in the short run to meet fluctuations in demand.[107] This is true not only of the market in general but also of individual customers, whose own production processes may be continuous and difficult to adjust. The matter is one of negotiating the respective streams of production and balancing the respective inventories. That makes a steel sales representative a "sort of foreign service officer, interpreting and adjusting the relationship between the customer and steel plant superintendents." Required is a thorough knowledge of both the mill's and the customer's facilities and technologies, and a "nice judgement in establishing a delicate balance between the limits of his own mill's scheduling and the patience and production plans of his customer."[108] With all integrated steelmakers facing the same rigidities, oversupply in the face of declining demand readily occurs. Marginal-cost considerations reinforce that behaviour, as will be seen shortly.

By the mid-1990s, a standard integrated operation was a blast furnace producing 4 million tonnes annually supplying two 300-tonne BOF furnaces.[109] To put that figure in perspective, Stelco's 2007 output was 3.8 mmt.[110] For comparison, the lower-middle ranks, comprising twenty-two of the world's top eighty steel producers, have outputs between 4 and 6 mmt.[111] That volume is still at the low end of estimates of minimum efficient scale for production using BOFs and the most modern blast furnaces.[112] The average Japanese steel plant has a capacity of 7 mmt, and a high-range estimate of minimum efficient scale for blast-furnace/BOF production is 10 mmt. As an example of high volume, the 2007 output of South Korea–based POSCO, the world's fourth-largest steelmaker, was 31.1 mmt.[113] At the finishing stage of production, continuous casting has reduced minimum efficient scale to 500,000 tonnes, making for ideal applications in minimills, whose electric-furnace technology has a minimum efficient scale ranging from 0.5 to 3.0 mmt.[114] Continuous casting can be adapted to the integrated mills' larger volumes as well, making the technology an improvement on both sides of the industry.[115]

High-volume technology may not be feasible to adopt. Earlier blast furnaces, for example, had to be shut down and relined every seven or eight years, requiring plants to have a second unit. New furnaces require relining only every fifteen years, and the interval is continuing to lengthen. Each furnace running at a minimum capacity of 3 million tonnes would require a mill output of 6 million tonnes. If market conditions dictated normal operations at 4 million tonnes, an expensive new furnace would be under-utilized, and the uncovered fixed costs would make it unaffordable.[116] Worse, given the large volumes involved, operating a second unit would risk over-supplying the market and putting downward pressure on prices. That, too, would raise the burden of fixed costs. Efficient utilization, not sheer scale, is the key.

Cost and large unit sizes complicate replacement decisions. One option is to keep the best furnaces in production and wait to see whether alternative technologies such as smelting reduction achieve the costs and volumes that warrant adoption. Investment in new blast furnaces is impractical when scrap prices are cheap because that cost favours minimills. These problems can be solved by eliminating blast furnaces and BOFs, purchasing raw steel from other producers, and concentrating on finishing it into high-end products. The beginnings of that pattern are appearing in the world's steel trade, as will be seen in Chapter 4.

Adopting minimill technology, which is efficient at small volumes, is a way of avoiding the rigidities and costs just seen. The major advantage is that minimills do not require the expensive and inherently high-volume stages of steelmaking, whose capital costs and high minimum efficient scale largely account for the integrated sector's rigidities and expenses. Saving the costs of steelmaking by relying on scrap would not solve the integrated mills' problems, however, because the scrap supply is not enough to support their production levels. As the minimill sector itself expands, its cost limits will hinge on scrap prices, and that will depend on scarcity. Given these supply and price constraints, blast furnaces – either on site or at remote locations supplying purchased raw steel – will likely remain the primary sources of hot metal for most steelmaking.[117] Because sheet steel is the integrated sector's product in greatest demand, flat-product mills will continue to be the core of large-volume operations. At the same time, the cost factors just seen make these massive facilities expensive, financially vulnerable, and dependent on adequate and stable prices.

Economies of Scale and Barriers to Entry

Scale economies occur when average cost – the cost of each unit produced – falls as output rises. Lower selling prices and greater profits are the potential gains, rewarding capital investments that increase volume. The rewards are greatest when fixed costs are already high. Indeed, "economies of scale are intimately bound up with sunk costs, i.e., irrecoverable investments."[118] In steel, according to one study, economies of scale are "sizeable for both open hearth and BOF plants." An earlier study of steel production found that doubling output from 0.5 to 1 million tonnes lowers total costs by 7.4 percent per tonne for open-hearth mills and by 8 percent for BOF mills.[119] One estimate places the output range for realizing scale economies in integrated mills at between 5 and 10 mmt.[120] According to a current study, adopting modern steelmaking technology yields "substantial" economies of scale, primarily because it sharply reduces labour costs. Those efficiencies make further improvements more affordable.[121]

Additional factors encourage large facilities. Complex and extensive production-flow processes usually require a single organizational structure and an integrated facility. In the steel industry, these requirements go beyond managing hot metal. The high indivisibility of production from blast furnace through continuous casting to final rolling requires "an incessant two-way flow of information ... to maximize energy efficiency and product quality."[122] This poses significant challenges related to scheduling and control. In one characterization, "the production process is sufficiently complex, involving closely articulated stages and continual adjustment and change, that the key lies in the coordination of expert systems. This is the case when a modern steel plant utilizes continuous casting methods to supply a number of finishing mills."[123]

Japanese steelmakers reaped the benefits of scale in the 1960s when they developed technology to raise the efficiency and output of blast furnaces and coke ovens, and when they built mills that made full use of those gains. Their pricing power made them formidable international competitors, and their profits supported further improvements. Some large European steelmakers also expanded their scale of production. American integrated mills, which previously enjoyed the greatest advantages of scale, began to lose them. Canadian integrated mills, meanwhile, had scale advantages not because of total volume but because of specialization, as will be seen later.

Scale economies can constitute a barrier to entry for new firms because incumbents enjoying those economies can produce and sell for less. High capital and start-up costs raise the barrier further. That has been the case in the integrated steel sector. New producers – chiefly in South Korea and Brazil – have entered the market over the past thirty years but have done so with the financial support of their governments. The scale economies enjoyed by larger producers have made small operations particularly expensive in comparison. The Chinese steel industry, which previous governments had intended to be based on small and local facilities, has sought to close marginal plants and consolidate larger ones and has used financing from government-owned banks. As will be seen in Chapter 4, these efforts have not been successful, and small mills have actually proliferated.

Along with scale, a key component of high entry barriers is specialized technology. Because large parts of mill facilities cannot be shifted from one product to a different one – from structural steel to galvanized sheet, for example – new investment is required to enter a different product line. Doing so confronts incumbent producers. In markets where they are significant players, the latter have "first mover" pricing advantages and can undercut new entrants.[124] That prospect, and the certainty of high fixed costs, makes entry unpromising and risky.

Entry barriers rise as production technology becomes specialized and automated.[125] This is true of manufacturing in general: "barriers to entry and exit for manufacturing firms appear to have risen markedly along with the growing importance of specialized technologies and as a function of the higher start-up costs and investments in physical capital associated with the general growth in the scale of production."[126] In the language of economics, the question is one of mobile and immobile factors of production. Low factor mobility signifies a relative inability to convert facilities from one product to another. This limitation accompanies high specialization and tall investment thresholds. It maintains barriers to entry and entrenches the position of incumbent firms. Those firms accordingly prefer existing conditions, and that attitude strongly affects their views of competition and trade policy, as will be seen in Chapter 3.

The factors that prevent entry to an industry may also prevent exit from it. The key is durable and specific facilities. If these cannot readily be converted to other uses, they compel a producer to stay with existing production even in the face of declining returns. That gives exit barriers the power, in Caves' and Porter's words, "to inflict persistently subnormal profits."[127] The more specific and expensive a firm's facilities are, the higher the exit barriers are, and the greater the prospective loss in the event of failure – especially if these facilities are large, immobile, and difficult to resell.[128] Large integrated steel facilities exemplify those conditions because of the cost barriers that discourage potential purchasers. The conditions worsen if the facilities in question are inefficient or obsolete. Standardized components incorporating current technology are available from global producers, and potential purchasers would naturally prefer them.

These obstacles induce integrated steelmakers to continue production in the face of declining returns. Reinforcing that pressure is the need to cover fixed costs – including those of the very facilities that prevent exit.[129] The key factor in deciding which facilities to close down is the firm's ability to tolerate losses. If these are tolerable, production will continue.[130] Adding to these conditions are the integrated steel industry's continuous production technology and the high cost of shutting down furnaces.[131] Together these conditions induce steel producers to persist in the face of low returns and discouraging prospects.

That is a recipe for price warfare. In the OECD's words, "one can generalize that the more significant are economies of scale, the more frequent and destructive price wars will tend to be. Prices can drop a long way before firms are no longer able to cover their variable costs. And with huge sunk costs, no firm wishes to permanently exit the market."[132] Other conditions encourage price wars as well, as will be seen shortly. Unwillingness to cut production also encourages price volatility. When demand falls, as was just seen, prices can sink rapidly. But when demand rises, steel prices "tend to soar to very high levels" because the industry is operating at a set level on the upside as well.[133] Price volatility, too, has important implications, as will also be seen.

For minimill operations, capital-cost barriers to entry are well below half those of the integrated sector. Exploiting those lower costs, minimills quickly expanded into the lower part of the product line. That they were able to do so is evidence that entry barriers can be ineffective protection in the face of a different and cheaper technology. The minimills' success in capturing eventually half of the North American market came at the expense of the integrated producers. Losing market share and revenue made the weight of their fixed costs even more burdensome.

Substitutability and Marginal Cost

Except for specialized top-end products, steel is a homogenous good manufactured according to common and specific global standards. Although the use of steel in individual products is governed by particular specifications, these are tailored to existing standards, and a steel consumer desiring something different must persuade a producer to develop it. That has, incidentally, occurred. Demand for thinner and higher-strength steels has driven advances in metallurgy. That has been particularly true in the automotive industry, which is the integrated mills' largest customer. Product development has been driven by the automakers' need to improve fuel efficiency by making lighter cars, which require thinner steel, and to improve crash safety, which requires stronger steel. Complex contours for automobile panels require steel that is highly formable. Driving development as well has been the steelmakers' need to prevent the automakers from substituting plastics, aluminum, and composites.[134] According to one industry analyst, "steelmakers claim that half of today's steel grades didn't exist ten years ago."[135] A government survey of the Canadian steel industry makes the same assertion: "over half of the steel products being used today did not exist a decade ago."[136]

Nonetheless, for the most ordinarily used items, one steelmaker's product is easily substituted by another's. Product homogeneity does eliminate one barrier to entry – an incumbent product's distinctive and preferred characteristics – but it also means that no supplier of standard commodity steel can depend on special qualities to attract and retain buyers. For their part, buyers may shop for price and for secondary features such as delivery and service. Prices can differ because there is no common mechanism for setting them. Prices for other primary metals are set on the London Metal Exchange, but with steel there are too many different varieties and grades. Each steel producer sets its own prices, making the process complex.[137]

Product homogeneity and highly similar producers make for a distinctive industry. An OECD study of industrial market structures classifies industries according to product differentiation and market fragmentation. Steel is located in the quadrant of low product differentiation and low fragmentation, meaning that the industry is composed of large and interactive firms producing standardized goods.[138] Interactivity and pricing will be discussed further in the next chapter. Because of these product characteristics and the industry's own highly standardized technology, steelmakers as well as their wares are "relatively undifferentiated."[139] Product homogeneity also means that steel producers are subject to the same market trends. Downturns that affect demand for one producer's structural steel or hot-rolled coil affect all other producers of those items. That shared fate provides one incentive for steel makers to collude, as will be seen shortly. Internationally, undifferentiated products affect trade. In the words of the chairman of the International Iron and Steel Institute (IISI), "ours is a bulk industry. Its products are interchangeable between one country and another."[140]

That raises the question of how to characterize the sheet products made by minimills. Using cheap, lower-grade scrap yields cost advantages, but this metal can produce only lower-grade sheet qualities. At particular quality grades, the products are undifferentiated, since integrated producers can produce the same grades, but between grades there are clear and important differences. To produce higher-grade sheet products with their technology, minimills must use more expensive scrap, augment it with new iron, and use more expensive double-shell equipment. When those factors are not present, their product is not substitutable with higher-end sheet products from integrated producers.

Steel production has low marginal costs. Those are the costs of producing an additional unit and are a standard means of calculating optimal levels of output. High marginal costs discourage an expansion of production, and low marginal costs encourage it. The issue is not merely theoretical. In the words of Roger Philips, CEO of the former minimill operator Ipsco, "for most of the integrated steel mills in the U.S., their marginal costs of producing a ton of extra cold-rolled material is the cost of the raw material and hardly anything else. So they're very low-cost on the margin. That's why when the market drops they go for price decreases: because so much of their cost is fixed."[141]

The same combination of costs also encourages firms to keep producing in the face of falling prices. "When fixed costs are high," Gary Clyde Hufbauer and Ben Goodrich point out, "it makes sense for struggling steel firms to continue running their plants so long as the marginal revenues from extra production at least cover variable costs."[142] Those costs, as was seen earlier, are normally relatively low, although soaring iron ore and coal prices in 2008 did allow steelmakers to trim production late that year in response to falling demand caused by the global financial crisis. Even so, that response cannot be continued in the face of long market downturns. Also encouraging continued production at lower prices are the inability to switch facilities to other products and the desire to avoid the direct and indirect costs of laying off workers.[143] The situation is highly interactive: when one producer cuts prices, the others, whose products are identical, must follow or lose market share. Revenues fall for everyone, and their ability to cover fixed costs declines. Overall, "economic logic at the firm level ensures depressed prices – and widespread operating losses – at the industry level."[144]

Rational behaviour in these circumstances is to form an agreement to manage levels of production. In the OECD's words, "if all leading firms are aware of these points ... it might happen that significant economies of scale and important sunk costs help foster coordinated interaction ... Firms should understandably wish to avoid destructive price competition."[145] That has indeed been a pattern: "Steel mills all over the world have responded to this tendency to price below total costs by implementing various methods to maintain price stability."[146] The history of the industry has been marked by collusive behaviour in both domestic and international markets. In the United States, an informal price leadership arrangement, with United States Steel setting the benchmark, prevailed for much of the last century. Signalling soon replaced explicit coordination, but the arrangement continued because it did provide a measure of stability and predictability.

Governments have sometimes co-operated. In Canada, as the steel industry expanded in the 1940s and 1950s, the government's administration of competition policy allowed the three steelmakers to apportion the market among themselves by specializing in different products.[147] Specialization enabled the steelmakers to delay entering new product lines until demand was sufficient to capture economies of scale.[148] In the European Community, as the steel industry there was adjusting to excess capacity, eliminating obsolete plant and consolidating in the 1970s and early 1980s, the European Commission introduced a formal set of quotas to regulate production.

Two developments in the 1960s ended private co-operation in the United States. The first was the entry of minimills, whose lower production costs gave them no incentive to participate in pricing arrangements. Instead, their incentive was to gain footholds in the market through price competition. This was less of a problem for the Canadian integrated producers because, as will be seen shortly, they were already specialized and could allow the minimills to supply other areas of the market. The second development was the growing presence of foreign steel exporters, whose lower prices quickly evoked charges of predation. These opened an episode in world trade in which established integrated producers in the United States, Canada, and the EU sought to use trade law defensively. Sharing an exposure to imports but also exporting to the United States, the Canadian integrated producers find themselves on both sides of the conflict. Chapter 4 will examine trade in steel, and Chapter 5 will examine the protective use of trade law and the securing of price stability through consolidation.

Inelastic Demand and Intermediate Goods

The demand for steel is inelastic: changes in price do not strongly affect the amount that can be sold.[149] That has two implications for steelmakers. The first is that no integrated producer whose share of the market is large enough for its actions to affect others can gain by raising output (recall what was noted earlier about oversupplying the market). The same amount will be consumed, and the adjustment will be made according to supply and demand: prices will fall for all producers. That result is a further incentive for steel producers to collude to set production levels. At the same time, the prospect of unilateral gain raises the incentive for individual producers to cheat.[150]

The second implication is that fluctuations in demand can have powerful effects on prices. "With demand so inelastic," write Barnett and Crandall, "and with large cyclical swings in demand, steel prices can become as volatile as some commodity prices."[151] The same volatility is true of profit margins.[152] Cutting prices in the face of falling demand is the natural move for each producer. That is so because the demand for any *individual* producer's steel is highly elastic: easy substitutability enables any producer to increase sales by offering discounts. That possibility leads to complex what-if calculations throughout the industry. We will see more about interactive pricing strategies in the next chapter. Exit barriers compound these forces by discouraging firms from abandoning loss-making activities.[153]

Steel is also a derived-demand product. As an intermediate good, it is subject to the demand for its end products. When demand for durable goods – automobiles and appliances, for example – falls during cyclical downturns in the economy, demand for steel falls as well. When that happens, given steel's price inelasticity and substitutability, along with the steel industry's high fixed costs, barriers to exit, and low marginal costs, producers will be driven to cut prices. As a result, "it is not surprising to see them regularly run up operating losses when their traditional clients are having to come to terms with cyclical falls in activity."[154]

In practice there is some variability. Minimill products, chiefly structural steel and bars, tend to be sold on a spot-market basis and to reflect pricing conditions immediately. Much of the production of cold-rolled sheet steel, one of the integrated producers' prime commodities, is sold in volume and on contract to large purchasers, such as the automakers. Those products are still subject to price competition, and the challengers have been foreign exporters.[155] Although the automobile and appliance manufacturers prefer dealing with local steel suppliers, imports do set price expectations. More general influences are also important. In the late 1990s, the world steel market faced oversupply and weak prices and the talk was of global over-capacity. After that, the rapidly growing domestic markets in China and India began devouring industrial commodities, including steel. Scarcer supply caused global steel prices to rise rapidly, along with the integrated producers' exports and revenues. The variability was striking: between 1980 and 2000, the price for a tonne of hot-rolled coil averaged $339. By 2002, because of oversupply, it had fallen to $250 – well below the loss point. Then by 2004, because of China's surging steel imports, it had risen to $400, producing record profits.[156] As of August 2008, reflecting higher ore and coal costs as well as demand, the price was $1,093.[157] But in mid-2008, the global financial crisis suddenly froze credit markets and frightened consumers, collapsing the demand for durable goods such as automobiles and appliances – industries that are two prime steel customers. In response, prices for hot-rolled coil fell to $600 by early winter. Adding to pricing pressure was excess steel supply – an indication that production cutbacks by the major producers had not been sufficient or that other producers had not followed suit. The number of producers in the industry made the latter possibility difficult to determine.

Steel does have price elasticity in the longer run. That elasticity derives from the demand for steel end-products, which have their own price elasticity. Steel price elasticity increases with the portion it represents of those goods' production costs. Demand for those goods, in turn, may be cyclical and over the longer term may depend on consumer preferences and willingness to pay. Steel price elasticity is also affected by patterns of substitution.[158] To reduce vehicle weight and improve fuel economy, the automobile industry has been substituting steel with aluminum and plastics since the 1970s.

The same substitution has been true in the economy more generally. By the end of the 1980s – a difficult and turbulent decade for steel producers – less than half the tonnes of steel per $1,000 of inflation-adjusted GNP were being consumed in the United States from both domestic and foreign producers than had been consumed in 1951. One cause of declining demand was the completion of major construction projects begun in the 1950s, notably expressways and other large infrastructure. Another was consolidation and line abandonment in the railway industry, which had once been one of the steel industry's biggest customers. Steel demand also declined because of indirect trade, with foreign steel content in manufactured imports lowering the consumption of domestic steel.[159] In contrast, steel production in Canada grew steadily through that period from an output valued at $305,734,984 in 1949 to $7,442,664,000 in 1984.[160]

What do these conditions mean for integrated steel producers? First, their high capital and fixed costs are a continuing burden that limits flexibility and choice. Second, steel's price volatility makes for unstable and unpredictable earnings and access to financing. Third, these conditions can be exacerbated by the incentive encountered by every producer to discount prices rather than scale back production. Magnifying these conditions is the fact that integrated steelmakers' technology requires continuous operation. These circumstances, when folded together, produce the worst of both worlds: very limited financial flexibility and high exposure to economic cycles and volatile prices. Failure to keep pace with technology exacerbates these difficulties.

Modernization

Introducing new technology to an industry makes existing technology comparatively less productive. Thus firms must decide whether to adopt it. One issue here is whether to invest retained earnings or raise funds on capital markets. The integrated steel industry's large, concentrated, and very expensive units of production make these decisions both complex and consequential. High capital costs and prospective advantages for new technology's adopters raise the stakes, especially in industries with a limited number of producers and significant economies of scale.[161] At the same time, if an existing plant is not fully depreciated and still functions well, abandoning it in favour of an expensive new plant is warranted only when the gains in efficiency or product quality outweigh the capital costs.[162] In addition, the opportunity costs – the value of the forgone alternatives – may be significant if there are closely competing requirements and if the funds are scarce or expensive. Opportunity costs may be especially high when the need for investment is spread across a large and outdated set of capital assets – a problem that is amplified by integrated steelmaking's indivisibilities of production. Because cost competitiveness is directly related to technology, failure to stay abreast of technological change – particularly when some in the industry are achieving great improvements in efficiency and product quality – carries grim consequences.

One reason to modernize is to lower production costs. Although high production costs are not a particular problem if they are the same throughout an industry, lower costs can provide pricing advantages, and gaining them is one incentive to be the first to modernize. Not doing the same leaves competitors in a stronger market position, especially if the products are highly substitutable. The laggard firms face serious trouble. As they lose market share, their operating rates and revenues decline, their profits available for investment shrink, and their ability to finance modernization on capital markets disappears. High fixed costs increase the peril. The same is true of adequate earnings generally. Firms with shallow reserves and poor credit are the first to fail under pricing pressure or during cyclical downturns. The result of a number of weakened or obsolescent firms is a sick industry. Conversely, high prices and profits provide the funds for modernization. That positive convergence, along with the abandonment of obsolete facilities and uncompetitive product lines, helps account for the general recovery in the North American and European steel industries after the crisis years of the late 1990s (an episode examined in Chapter 5). That recovery is a reminder that conditions can change.

As has been seen, steel technology is specific to particular parts of production. When it is assembled as a set of closely connected components in an integrated facility, steel technology has high indivisibilities. An obvious modernization strategy follows: replace an entire facility, or build a new one at a brand new site. Only a few entirely new greenfield facilities, however, have been built in the wealthy industrial countries. Japan built them throughout the 1960s, but it was positioning itself for an expanding international market. That was part of an explicitly export-based strategy based on large economies of scale, modern technology, and seaport-based production facilities to take advantage of the ocean supply of iron ore and coal.[163] Since then, any new plants have been built in newly industrializing countries – principally Brazil and South Korea – although a greenfield facility was built in Canada, as will be seen shortly. The barrier is cost: approximately $10 billion for a completely new integrated plant. In comparison, a new minimill can be built for $350 million.

In addition to that investment hurdle is the prospect that returns may be inadequate. Fully modernizing an integrated mill requires a return of at least 12 percent. Even in good years, returns below 8 percent are a more normal pattern, and for North American steelmakers average returns above 8 percent occurred only twice between 1998 and 2003.[164] Another risk is unexpected market developments or economic downturns. For a cyclical industry such as steel, these fluctuations may leave a new facility uncompleted, compounding costs and the prospects of failure.[165] Existing weakness magnifies the problem: the steelmakers most needing to modernize are the ones least able to do so.

Incremental modernization is less daunting. One attraction of this approach is that it retains existing plant while avoiding the risks just described.[166] When modernization proceeds in well-coordinated phases, no older facility is "uniformly outdated."[167] The design and scheduling task is to maintain an ongoing balance among the connected stages of production. Modernization "has tended to be piecemeal [with] selected new equipment often being retrofitted with obsolete equipment."[168] The process resembles "the forward motion of a caterpillar. First one segment is extended, then the other attached segments are moved up." Because production stages are so closely interdependent, limitations at any one stage affect all the others. The result is that, given the high indivisibilities involved, the various components of an integrated steel facility are seldom a balanced system.[169] Because these systems are less integrated and streamlined than completely new installations, they cannot achieve the same process efficiencies; at worst they may merely amount to "splicing new technology onto old."[170] The imbalances can increase when improvements in some sectors are accompanied by the closing of obsolete facilities in others.

Reasons for retaining facilities multiply with external economies – that is, with localized conditions that support particular kinds of production. The most obvious such conditions are specialized labour and supply markets; indirect ones are communities of expertise and ancillary service. External economies develop around large individual facilities and are especially abundant in places where there are several firms in the same industry. External economies constitute an incentive to stay put. There are offsetting disadvantages, however. If facilities are large as well as outdated, the benefits of incremental improvements and retrofits may disappear under high costs and extensive inefficiency. This is a particular disadvantage if competing producers have completely new facilities. If manufacturing activity is shifting to different regions, staying in an existing location also means conceding the new markets to other producers because of the relatively high costs of overland transportation. In North America, this created opportunities for minimills and exporters. At the same time, integrated facilities are such large capital investments that relocation may not be an option, particularly if profits are under pressure. These opposed forces have directly affected the steel industry's global development over the past twenty years (see Chapter 4).

There is also the question of appropriate volume and scale. New technology in steelmaking generally raises capacity, yielding economies of scale and providing significant cost and pricing advantages, as was just seen.[171] These advantages did much to fuel Japan's advance in the international steel trade, so emulation by others would seem to follow naturally. However, increasing scale is not a benefit if it oversupplies the market, depresses prices, and raises the burden of fixed costs. That is why there is concern that individual countries, as they industrialize, will expand steel production not only to meet expected domestic demand but also in anticipation of export earnings. The prospect of countries pursuing these objectives unilaterally raises the spectre of global oversupply, and China's vigorous expansion is a particular worry. As will be seen in Chapter 5, avoiding oversupply through coordinated investment and production is one of the arguments for global consolidation. China's steel industry will be examined in Chapter 4.

A more immediate solution is to couple modernization with specialization – a strategy with applications in a number of industries besides steel. In addition, there is restructuring, which can include acquiring other firms, merging with them, or forming alliances with them. It is often possible to pursue modernization, specialization, and restructuring at the same time.[172] Mergers and alliances have been part of the globalization of steel; they appeared first between Japanese and American steelmakers and more recently among European, Russian, and North American firms. Mittal Steel, which began its rise to current pre-eminence by purchasing and consolidating inefficient steel producers in Asia (often state-owned ones), has gone on to make acquisitions in North America and Europe and intends to pursue the strategy worldwide. The purpose is to achieve economies of scale by combining efficient components into new production systems tailored to particular markets. The global dimensions of this strategy are considered in Chapter 6.

One option is to concentrate expensive and high-volume blast furnaces and BOFs in one or a few sites and select rolling mills and special product lines according to the most optimal combinations of demand and productive efficiency. Where possible, these should be oriented to the requirements of principal clientele.[173] In contrast to the hot-metal stage, scale in the finishing stages should be limited to the "minimum necessary to gain adequate economies."[174] This combination reduces the level of capital costs and makes it possible to maintain high operating rates. Facilities that do not fit these requirements are retired or sold. Smaller producers may need to become highly specialized either by directly serving specific regions and industries or by becoming a "dedicated niche provider of high-value, non-standard steel products that can withstand the costs of travel to global markets."[175] One medium-sized producer that has done this quite successfully is Austria's Voestalpine.

In the mid-1980s, American integrated steelmakers faced two new competitors – domestic minimills and foreign exporters – both of which enjoyed strong advantages at the lower end of the product line. Some of the foreign exporters, notably South Korea, also had much newer and more efficient facilities. Lacking the resources for comprehensive modernization, American producers specialized and downsized. They abandoned low-margin product lines, modernized the remaining facilities, and cut capacity. The scale of reduction meant forgoing the new and highly efficient blast furnaces developed in Japan because they exceeded plant size. The producers concentrated on higher-end specialty steels and flat-rolled products, particularly coated sheet, where demand was actually growing. To meet sheet-steel demand in the automobile industry, supply capital, and circumvent trade restrictions, Japanese steelmakers in the 1980s entered joint ventures with several American integrated producers. Between 1989 and 1995, the integrated producers added 5.8 mmt of flat-rolled capacity, representing an increase of 20 percent over 1988. Since then they have added another 17 mmt.[176]

The experience of Canadian and American integrated steelmakers with modernization and specialization illustrates these considerations, although in different sequences.

Strategic Choices

Because tariffs on steel were relatively low, steel did not follow the pattern of other industries, in which American firms established Canadian subsidiaries or acquired Canadian firms. The Canadian steelmakers' strongest advantage was their domestic supply of iron ore, but that was offset by the cost of pre-processing and transporting it, by low labour productivity, and by high costs of capital. Even so, in the 1950s the Canadian government, viewing resource-based industries as Canada's natural advantage in the world economy, decided to encourage the steel industry.[177] The crucial choice was between scale and high operating rates.

The latter was chosen. As was noted earlier, the Canadian government allowed the three major steel producers – Algoma, Dofasco, and Stelco – to specialize in particular product lines, although Stelco was closest to American practice in its breadth of offerings. With the government's acquiescence, the steelmakers staked out areas of domestic market dominance, thus overcoming the disadvantage of a small domestic market and gaining economies of scale in their particular niches. Operating rates were accordingly high. Between 1965 and 1975, for example, the three steelmakers operated at between 78.5 percent of capacity and 99 percent. The former figure was set during a steel strike in 1969. Over the longer period of 1960 to 1980, the Canadian steelmakers' average operating rate was second only to Japan's.[178] Translated into costs and competitiveness, higher-capacity utilization gave Canadian steelmakers a 20.9 percent advantage, between 1955 and 1970, over American steelmakers.[179] After-tax rates of return between 1961 and 1974 averaged 9 percent, compared to 2.5 to 7 percent for American steelmakers. These rates enabled Canada's big three to finance much of their growth internally.[180] Each steelmaker's dominance in its particular parts of the market lowered investment risks; this allowed the steelmakers to modernize production. To offset the relatively high costs of capital, government tax provisions allowed very rapid depreciation.

Imports were treated both as backstop supply and as sources of specialty steels for which the domestic market was too small to yield scale economies. That policy limited surplus capacity to cover periods of peak demand; it also allowed steelmakers to concentrate on high-volume products. Imports also provided a buffer for Canadian steelmakers to expand production, and as they did so, imports shrank. As their advantages grew, the steelmakers began exporting to the United States.[181] Exports increased along with production, and in 1976 Canada became a net steel exporter.[182] At a time when steelmakers in the United States and Europe were regarding greenfield plants as prohibitively expensive, Stelco built an entirely new facility at Nanticoke, Ontario, to which it devoted two-thirds of all its investment in the 1970s.[183] One reason for taking on this huge project was to avoid piecemeal modernization at its existing facility in Hamilton. Throughout the 1980s, the Canadian steelmakers' focus was on obtaining the best technology available.[184] Established patterns of specialization meant also that minimills were not the same threat to Canadian as they were to American steelmakers.

The opposite pattern characterized the American integrated steel industry. In the 1950s, it invested heavily in additional capacity; however, it did so by adding open-hearth furnaces. These raised volume but not efficiency. They became a liability in the 1960s, when the growth of steel demand fell short of industry expectations and when Japan began gaining significant export advantages from modernization and scale. The American steelmakers had incurred these large sunk costs just as BOF technology was becoming available. With weak demand and revenue, the steelmakers lagged behind their Japanese and European counterparts in adopting BOF technology. Worse, observing the Japanese steelmakers' gains from scale economies, the American steelmakers sought to maintain capacity despite slackening markets.[185] The resulting financial weakness made it even more difficult to modernize further. The difficulty became severe in the 1970s, just when their counterparts in Europe and Japan were installing continuous casting.

Financial weakness and large, obsolescent facilities dictated the choice of piecemeal modernization, but by spreading scarce capital over extensive facilities, the steelmakers were unable to realize strong returns. Adding to these difficulties were large wage-rate increases, which raised hourly pay by 68 percent between 1972 and 1977 in return for a 3 percent gain in productivity.[186] The union's ability to negotiate those increases actually provided a *dis*incentive to modernize: "The expected excess return was zero since the unions would appropriate the rents, ex-post."[187] These cumulating disadvantages became critical with the advent of low-cost minimills and the rise of foreign imports.

During the 1970s, while it was prospering, the Canadian steel industry began expanding volume. That move, which the government encouraged, was seen by some at the time as a strategic mistake.[188] High operating rates had come from specialization in a moderate-sized market, and higher volume would dissipate those benefits. A large wage settlement in 1981 further reduced the industry's advantages. Higher volume meant relying increasingly on exports to the United States, although declining cost advantages, which began as American steelmakers started rationalizing, forced a growing dependence on favourable currency exchange rates.[189] Depending on exchange rates is risky for any steel producer, particularly when unusually favourable rates are taken as a basis for longer-term estimates. In addition, the export benefit of a cheap Canadian dollar is offset by the pricing of coal and iron ore in American dollars.[190] The advantage of favourable exchange rates ended in 2003 and 2004, when the Canadian dollar rose from US$0.63 to US$0.83; by 2008 it had achieved parity.[191] At the same time, however, the American market has very high steel prices, enabling Canadian producers to sell one-third of their output there.

More recently the Canadian iron and steel industry has focused on modernization, spending $7.8 billion between 1980 and 1995 and $4.4 billion between 1990 and 1998.[192] At an average capital outlay of US$23.60 per tonne, the Canadian integrated producers slightly exceeded the average of their American counterparts, although both lagged behind producers in all other major steel-producing countries. The Asian producers' high figures reflected capacity expansion, while those of the North American producers reflected modernization.[193] Because Canadian steelmakers had been quicker than American ones to adopt new technology, their labour costs were lower until 1987, when the American producers' massive restructuring began delivering improved efficiency. Canadian labour costs again fell slightly below American ones in 1992, partly because of changes in the currency exchange rate.[194] Because recent investment had been directed to increasing efficiency, capacity growth was modest, and when domestic steel demand rose in the early 1990s, production was insufficient. In 1994, Canada, which had been a net steel exporter since 1976, again became a net steel importer.[195]

Even though domestic steel demand and steel shipments by Canadian producers in 1997 were the highest in a quarter-century, the following years saw deteriorating conditions for two of the three integrated producers. Algoma Steel, the smallest and weakest of the three, had already entered bankruptcy in 1990. Its basic disadvantage was locational. Situated in Sault Ste. Marie, Ontario, the mill's original plan in 1898 was based on faulty estimates of local iron ore and fuel quality. The result was a cost structure that included expensive transportation for both raw materials and finished steel.[196] In the 1980s, an aging plant and falling demand for its lower-end products compounded Algoma's problems. Under reorganization it moved upmarket, specializing in cold-rolled sheet. It was one of the first steelmakers to install a thin-strip caster.

Although Algoma, along with Stelco and Dofasco, had higher levels of profitability than all but one of the American integrated producers, AK Steel, Algoma declared bankruptcy a second time in 2001, citing the effects of imports. In 2004, Stelco declared bankruptcy.[197] Stelco's disadvantage was that it had restructured later than its competitors and stayed in lower-end product lines where imports and minimills were competitive. Additional disadvantages were higher labour and pension costs than those of restructured and surviving American steel producers, and higher export prices due to a rising Canadian dollar.[198] An analysis of the Canadian steel industry in 1991 predicted that demand in the American market for Canadian steel would shift to the upper end of the product line.[199] By specializing there Dofasco remained viable through the worst of the world steel downturn of the late 1990s – so much so that Dominion Bond Rating Service ranked Dofasco, along with the American minimill Nucor, as North America's two premier steelmakers.[200]

Instead of expanding volume, American steelmakers in the 1980s began closing obsolete facilities. By 2000 they had removed some 35 percent of the industry's capacity.[201] The industry trimmed payrolls, and employment declined by 75 percent between 1970 and 1995.[202] Once the producers did begin installing continuous casting, its share of production grew rapidly, from 39.6 percent in 1984 to 61.3 percent in 1988. By 2007, continuously cast output had risen further, to 96.7 percent. Canada's portion was 99.8 percent.[203] The industry gradually began to concentrate on product lines where its advantages were strongest and to invest accordingly, with total investment totalling more than $50 billion by 1998. The combination of investment and labour cutbacks doubled productivity.[204] United States Steel, for example, concentrated on flat products and exited the market for bars, rods, and structural steel. Two-thirds of the company's plants were closed, and capacity, along with the labour force, was reduced by half.[205] To increase its production of flat products, it purchased National Steel in 2003, adding seven million tons of capacity.[206] The integrated steelmakers' product quality improved as well. Responding to the automobile industry's demand for better steel quality and dependable delivery, the steelmakers achieved a 90 percent reduction in the automakers' rejection rates. And in contrast to the experience of the steel industry in the EU, the American industry's recovery was accomplished with the industry's own resources, with the major government interventions having been the imposition of a set of voluntary export restraints during the 1980s and a special steel tariff in 2002.[207] (Trade protection will be discussed further in Chapter 5.) Despite these measures, thirty-three American steel producers have declared bankruptcy since 1997.

Failure and Salvage

Much of the problem lay in the unfavourable combination of high capital requirements and unstable earnings. The risk of financial overextension is very real. For cyclical industries such as steel, net-debt-to-capitalization ratios must not exceed 40 percent to avoid failure during troughs in the cycle. American integrated steel producers exceeded that limit every year between 1998 and 2003, with the best year being 1998 at an average of 50 percent and the worst year being 2003 with an average of 75 percent. In a stronger position, the Canadian producers' worst year was 2001 at 48 percent, and their best year was 2003 at 35 percent. The least indebted North American producers were Dofasco and Nucor.[208] For the rest, their high debt levels weakened their ability to withstand periods of excess supply and strong price competition and, over the longer term, to afford the improvements to stay abreast of more efficient producers. As one index of weakness, the bonds of eleven of the fourteen steel producers surveyed by Standard and Poor's in 2001 – one of the trough years of steel prices – were rated as junk.[209]

The steelmakers' investments did leave modern and usable production assets around the industry, inviting salvage and recombination into new firms. Mergers and acquisitions are indeed a way to reduce losses, reassemble the most efficient elements of production, and specialize in the most advantageous product lines. It is possible, in other words, to combine specialization with improved scale economies and more streamlined and efficient production.[210] As one index of consolidation, in 2001, over 80 percent of flat-steel products were produced by the top ten American steelmakers; by 2004, almost 75 percent were being produced by the top three.[211]

The largest mergers have been in Europe among previously state-owned steelmakers. Corus, for example, was formed in 1999 when privatized British Steel merged with Dutch steelmaker Koninklijke Hoogovens to become the world's fifth-largest integrated producer. Luxembourg-based Arcelor, until recently the world's largest producer, arose in 2002 from a three-way merger of France's Usinor SA, Spain's Aceralia Corporacion Siderurgica SA, and Luxembourg's Arbed SA. Netherlands-based Mittal Steel was itself the product of a 2005 merger between two family holdings (Ispat International NV and LNM Holdings NV) and International Steel Group in the United States. In 2006, Mittal acquired Arcelor; the result was the world's largest steel firm.

International Steel Group (ISG) had been formed in 2001 when investor Wilbur Ross purchased the assets of LTV Corporation. Reorganizing as ISG, Ross then acquired the assets of bankrupt Bethlehem Steel, Acme Steel, and Weirton Steel. LTV itself was a consolidated entity, having previously acquired Jones and Laughlin Steel and Republic Steel, and had been in bankruptcy twice in the previous twenty years. By the time of the merger with Mittal, ISG had reorganized those assets into a viable business, providing Mittal, which had already acquired Inland Steel in 1998, with a dominant position in the American steel industry.[212] In 2005, the same prospect induced the Russian-based steelmaker, OAO Severstal, to submit a bid for Stelco, having just purchased the bankrupt American steelmaker Rouge Industries, formerly the steelmaking unit of the Ford Motor Company. An additional motive was to gain entry to the North American automobile industry, which was a primary customer of both Rouge and Stelco. The strategy of both ISG and Mittal shows that properly targeted acquisitions can revive salvageable parts of even deeply troubled industries.

At the same time, prospective merger partners and buyers face the costs of retirement benefit obligations. By the time Bethlehem Steel, formerly America's second-largest integrated producer, declared bankruptcy in 2001, its single largest financial outlay was its pension fund. Similarly, Stelco had more than $1 billion in uncovered pension liabilities, and funding them would draw heavily on any new financing and revenue. These burdens, which grow as producers modernize and reduce their workforces, offset restructuring prospects as well as the appeal of attractive assets.[213] Bankruptcy is one way of removing them. By buying LTV at a bankruptcy auction, Wilbur Ross was able to transfer the company's $14 billion pension obligations to the federal government's Pension Benefit Guaranty Corporation and shed another $5.9 billion in retiree health-care costs – a move that helped make the company very attractive to Mittal but that also left retirees with significantly reduced benefits.[214] Another avenue is government assistance. Consolidation in Europe was made easier because the governments helped ease the effects of shutdowns and layoffs.[215] The Ontario government, as part of its rescue package for Algoma, took over the company's pension obligations to retirees.[216]

Conclusion

These facts evoke a sense of freedom and dread. Fully able to act according to their best estimates and interests, steelmakers enjoy wide latitude, and as autonomous entities they are free to decide. At the same time, as their governments have withdrawn supports and protections and opened markets to global forces, their milieu has become more unstable and unpredictable. Each steelmaker's situation is contingent not only on variable and widely fluctuating conditions among suppliers and customers but also on the behaviour of fellow producers. Dread – an awareness of fateful uncertainty – arises from the limited ability of any individual producer to affect these conditions and from knowing that fellow producers may calculate their survival in ways that offer no protective community. The backdrop of today's active and prosperous milieu is an expanse of failed firms and closed facilities. It is all too easy to join them.

The reasons are rooted in basic economics. Steelmaking as an industry has expensive and immobile capital, produces a highly substitutable product, is subject to overproduction and falling prices, and is fully open to severe cyclical forces. The heavy cost burden and the operating rates that burden imposes limit short-term flexibility and magnify the impact of weak prices. Longer-term adaptability is limited by the costs of modernization, the size and scale of production facilities, and the difficulty of obtaining dependable financing. Soaring world demand after 2001 supported seven years of good earnings and solid finances, but cycles can reverse. Such a reversal occurred rapidly in mid-2008, as the global financial crisis caused demand for steel-containing goods to plummet.

When competition prevails, as it has over the last thirty years, and when individual producers are large enough to affect the others, pricing calculations become interactive. The strategic possibilities of that behaviour fascinate economists, and their models provide a usefully concise way of considering steelmakers as players in a market, as Chapter 3 will show. The rise of large foreign steel producers makes the market international, and Chapter 4 explains how that has happened. Steel tariff barriers have fallen under successive phases of world trade liberalization; one result is that European and Asian firms now dominate world production. At the same time, the basic economic factors just seen affect the world's steelmakers impartially. The fact that globally dispersed producers are subject to the same forces and incentives makes those economic factors important to know.

When market conditions are troubled, as they have been periodically since the 1960s, steelmakers in Europe as well as North America have blamed foreign imports. Chapter 5 takes up the legal and policy implications of dumping. It also considers two alternative strategies: industry consolidation, and forging interdependent ties with major customers. Chapter 6 considers the ways a globalized steel industry might evolve.

3
Prices, Preferences, and Strategy

The way steel is sold is influenced by the basic economic features just seen. Producers in the integrated steel sector supply large market segments, are small in number, and are affected by one another's actions. They share high fixed costs, a vulnerability to swings in demand, and an incentive to discount prices before cutting production. They have a common interest in stable and agreeable prices, and although organizing collusion is illegal, informal co-operation within an understood price and output range is possible. At the same time, each producer knows that its cost structure allows it to capture market share by cutting prices and raising output. These opposed incentives have long characterized integrated steel, as has relative price stability.

Stability ends when new competitors seek market share for themselves. In steel, these competitors have been domestic minimills and foreign exporters. Both expanded through price competition, and their entry came just as the informal understandings among the incumbent producers were breaking down. Although discounting by a domestic producer is not illegal unless it can be proved to be predatory, discounting by a foreign producer invites charges of dumping. Forced to accommodate domestic minimills, the integrated steel producers directed their unfair pricing complaints at exporters.

Dumping – selling goods in export markets below home prices or below the exporter's cost of production – has long interested economists, and a focus of more recent research has been export-pricing and market-share strategies. Many of these strategies hinge on price discrimination – selling a good at different prices in home and export markets. When a domestic producer can prove dumping and injury, remedy is available under trade law. At issue are when that intervention is justified and what is to be protected – the producer or competition. The question interests political economists, and one focus of their research is how trade policy preferences are affected by the characteristics of particular industries.[1] A related interest is the general disposition of governments. This chapter will briefly explain pricing incentives in steel and the economics literature's view of dumping and strategies for capturing market share. The chapter will conclude with international political economy's views of the motives for seeking protection and the factors affecting governments' inclination to intervene.

Life in an Oligopoly

The integrated steel industry fits the standard profile of an oligopoly. Economies of scale enable a few firms not only to supply the market but also to affect the price. The importance of that fact can be seen when we compare it with what happens in perfectly competitive markets. The latter are characterized by many small producers who must accept a prevailing price as given and who know that their actions will change neither that price nor the behaviour of the other producers. In economic theory, such markets are self-regulating and can yield outcomes in equilibrium. In oligopolistic markets, producers can affect both price and one another's behaviour because their share of the market is significant. Because of their ability to affect outcomes, there is no single equilibrium in oligopolies; rather, there are multiple ones, as well as multiple incentives and degrees of influence.[2] These conditions lend themselves to strategic calculations wherein each participant must assume that all desire the best returns for themselves but do not wish others to get more. A tension prevails between individual gain seeking and co-operation. One way of securing mutually positive returns is by co-operating to limit production and price competition.[3]

Basic conditions in the integrated steel industry, however, encourage gain seeking. High entry and exit barriers, together with high fixed costs and low marginal costs, induce producers to meet falling demand by cutting prices. That is so because high fixed costs must be covered and low marginal costs make it inexpensive to adjust quantity.[4] An easy option is simply to continue at current volume. If producers believe that downturns will be temporary, they will prefer to carry on production at lower prices; if prices do not fall below variable costs, there is the prospect of covering some fixed costs. These incentives are common to cyclical industries.[5]

Although such gain seeking – or more accurately, loss limiting – behaviour is possible in an oligopoly, there may be limits. One limit is each firm's prudent choice not to overinvest in capacity. When all practise restraint, total supply is sufficiently contained to keep prices from falling to ruinous levels even during a recession.[6] The previous chapter showed that overexpansion has indeed been a consideration in the industry's investment choices; in this regard, the next two chapters will show that there is much concern about overexpansion in China. Producers may, to be sure, invest according to individual and not collective incentives, and limiting capacity to maintain a price floor requires an appreciation of the industry as an interdependent entity.

The most explicitly co-operative solution is collusion. Oligopolists know that by regulating output, they can supply the market at a price above marginal cost, with the difference representing pure profit. High entry barriers keep out new competitors and make it possible to raise prices by restricting supply.[7] Formal and explicit coordination is illegal, but price leadership and trade associations provide alternatives. Under a system of price leadership, a dominant firm announces prices and the other members follow suit. Success depends on the dominant firm selecting an agreeable price and the others adhering to it.[8] When that arrangement prevailed in the American steel market, the price leader was United States Steel and the benchmark price was f.o.b. Pittsburgh.

The need for price leadership does not arise in segmented markets because the participants do not compete directly. As was seen in the previous chapter, Canadian steelmakers followed that strategy in the 1950s and 1960s when they established specialized domestic market shares. Trade associations facilitate informal co-operation by providing members with information about costs, production levels, prices, and by conducting market analyses to forecast demand. Members can use that information as a basis for decisions about pricing and capacity and in order to anticipate the behaviour of the other members.[9] Information about output and market conditions is collected by the Canadian Steel Producers' Association and by the American Iron and Steel Institute. Data can be found on their websites at http:///www. canadiansteel.ca/ and at http://www.steel.org/.

In an oligopoly, co-operation entails producing and pricing at expected levels, and defection entails raising supply or cutting prices. The temptation to defect grows with excess capacity, high fixed costs, a substitutable product, and a price pegged well above marginal cost. According to the OECD, "once price has been raised significantly above marginal cost, each oligopolist will have an incentive to shade its prices, especially if it believes this will escape detection and retaliation."[10] The size of the potential gain increases with the other firms' difficulty in detecting defections and organizing retaliation. Uncertainty adds further encouragement. Other firms may not be able to tell whether prices are falling because demand is lower or because a member is cheating.[11] They also may not know one another's actual production capacity and potential for market disruption. That is true in the integrated steel industry: "The real capacity of steel plants is exceedingly difficult to determine and often is a closely guarded commercial secret."[12]

Even so, because members of an oligopoly are all large enough to affect the market, any member contemplating defection must take into account the responses of the others. An obvious response to a unilateral price cut is for the others to lower prices as well. This leaves everyone worse off, but it also neutralizes the defector's gain. The likelihood of this happening is a deterrent. A Nash equilibrium (formulated by mathematician John Nash in 1950) prevails when no firm, anticipating the others' responses, can move unilaterally to improve its position. The outcome is stable when it becomes a "set of self-enforcing actions from which no firm would unilaterally wish to deviate."[13] Restraint may arise more informally from individual judgment and experience, and may be particularly likely to develop in an industry with established common understandings.[14] Because these understandings are tacit, prospective, and deterrent, they fall below the threshold of collusion. Other conditions reinforce that stability. Particularly important is the industry's concentration in a small number of producers and their willingness to hold large inventories to manage market supply. Offsetting that influence, however, is the underlying condition of fixed costs. When these are high and profit levels are low, "fragile pricing discipline" may be the cause.[15] Otherwise, price coordination would deliver more ample returns.

These conditions need not prevail if products are differentiated even slightly. When that is the case, the need for an industry to coordinate diminishes, even if producers enjoy economies of scale and the market is not fully competitive. Differences allow each firm to have a "monopoly on its own distinctive product." If enough buyers are willing to accept a different but highly similar product, enough sales will be diverted to eliminate excess profits, and the potential market latitude will allow each firm to operate without need of collusion.[16] Identical products, on the other hand, mean that no producer can claim an advantage except on price. As was seen in the previous chapter, commodity steel is produced according to uniform international standards. That fact supports incentives among producers to coordinate output and regulate prices.

Because behaviour in oligopolies is based on calculation and interaction, it is inherently strategic. It lends itself readily to game-theoretic modelling, for which the economics subfield of industrial organization has produced a large and sophisticated literature. These treatments quickly become technical, raising the question whether the same elaborate calculations actually guide firms in the real world. In fact, the decisions are likely to be shaped more practically by expectations of potential gain weighed against prospective retaliation and assessed in light of experience. "Not only may firms be unable, for reasons of bounded rationality, to work through the complex mathematics of these models, they may not have to because the answer is to them so obvious."[17] These kinds of understandings may produce considerable stability over time. So also may the prospect of long-term interdependence.[18] That stability may not be completely undesirable. Industries that are too open to new entrants may drive incumbents below their levels of efficiency, raise costs and prices, and reduce aggregate welfare.[19]

Outside Disturbance

Oligopolies can be thought of as integrated communities of producers joined by common risks and benefits and maintained by common understandings and practices. The system breaks down when producers from outside the community enter the market. Even when they charge the prevailing price, their presence undercuts an oligopoly because of increased supply. The effect magnifies when they pursue market share by discounting prices. Minimills, as the previous chapter showed, have lower cost structures than do integrated producers and offer competing products at the lower end of the line. When they began entering the market they had no reason to be part of pricing arrangements among the integrated producers and quickly began claiming market share for themselves. As was also seen, their impact at the time was less in the Canadian steel market because of the integrated producers' greater specialization. In the American market, however, the minimills entered just when the integrated producers were coping with falling demand, carrying large costs, and operating aging facilities. The American minimills' output doubled between 1974 and 1985 while that of the integrated producers fell by one-third.[20] Because of the integrated producers' shaky financial position, retaliating by matching the discounts was not a strong option. The breakdown of price co-operation left them on their own in a destabilized market.

While that was happening, imports were becoming a presence in both markets. Like the minimills, the exporters could gain market share by offering discounts. When they did so, the integrated firms accused them of dumping, a violation of trade law. Under the WTO's standards, dumping occurs when firms sell in foreign markets at prices below those in their home market or below the cost of production. Industrial countries, including Canada and the United States, have long had anti-dumping statutes on their books according to which tariffs are available to eliminate discount margins on imports that are shown to be causing injury to domestic producers. Canada's anti-dumping law dates from 1904 and the United States' from 1921. Chapter 5 takes up the use of trade law as a survival strategy. In this chapter the question is discounting itself. If domestic producers have good reason to prevent destructive price wars, why would an exporter cut prices below its own domestic market levels or, more drastically, below the cost of production?

The basic issue is price discrimination – charging different prices to different buyers. Price discrimination is common in domestic commerce. It can be seen openly in student and senior discounts, and it is present less visibly whenever buyers must haggle with a dealer. The dealer has an unstated floor price and seeks to extract the maximum surplus. Determined bargainers may reduce the surplus and pay less, but others will pay more.[21] As was just seen, oligopolies earn extra profit by agreeing to production levels that set price above marginal cost. When that profit margin exists, and when additional buyers can be attracted by charging them less, they can be supplied at a discount. All the other buyers still pay the regular price, but the extra buyers get a bonus. In terms of economic welfare, this price discrimination can be beneficial if it makes available goods that otherwise would not be supplied.[22]

The same incentives prevail internationally among oligopolistic industries selling homogenous products. Trade is actually *encouraged* because firms hope to capture additional sales in foreign markets at discounted prices.[23] The calculations are based on price discrimination in which different buyers – domestic and foreign – are charged different prices. In the words of James Brander and Paul Krugman, the key element is a "segmented markets perception: each firm perceives each country as a separate market and makes distinct quantity decisions ... and chooses the profit-maximizing quantity for each country separately."[24] For this strategy to work, transportation costs and other disincentives must prevent the discounted goods being sent back to their home market and resold.[25]

The model, based on the economic theory of Cournot competition, assumes that firms make their production choices by quantity and take the other firms' quantity as a given. The exporting firm assumes that its domestic counterparts have already made their output decisions and that by offering a discount it can capture some of those producers' sales. Having already committed themselves to particular outputs, the domestic counterparts are faced with a choice: lose sales, cut prices, or scale back output. Brander and Krugman go further and assert that the incentives can operate in both directions, with firms in each country seeking extra sales by discounting. This process they term *reciprocal dumping*.

The assumption that output choices are made by quantity is not unreasonable for integrated steel producers, who must cope with rigidities in their production process, although some economists believe that ignoring prices overlooks important parts of firms' strategic choices and that in practice firms reckon both price and quantity.[26] Even so, the central consideration remains: when products are homogenous, when demand is inelastic, and when imports are discounted, buyers attracted to those discounted products will be lost to domestic producers. These considerations need not prevail if there is some product differentiation. Some buyers will be induced to switch to a different and cheaper imported product, but others will remain with the higher-priced domestic one.[27] With commodity-steel products, however, that redeeming latitude is not present.

Moreover, economic interdependence and trade fashion the world's demand for steel into a common cycle, with fluctuations affecting all producers even though conditions in their home markets may be different. When that difference exists, steel producers must adjust to two sets of conditions.[28] One adjustment in the face of falling home-market demand is exporting at discount. Doing so when overall world demand is also falling, however, magnifies the dislocation for domestic producers in those export markets.

Another steel industry constraint is high fixed costs. The higher they are, the more willing a firm will be to discount to help cover them, and the longer it will be willing to persist in doing so.[29] As was seen in detail in Chapter 2, high fixed costs are one of the integrated steel industry's key facts of life. Continuous production processes, heavy sunk costs in expensive plant, and production concentrated in very large facilities also encourage discounting.[30] So also do low costs of processing – represented by variable and marginal costs – and marketing. Those conditions, too, are basic to integrated steel. When they are present, and when there is a surplus on the world market, there is "always the temptation to dump steel ... to recoup variable costs and sustain a stable cash flow. Thus the liquidation of steel inventories at dumping prices (prices lower than average total cost) can provide a means of short-run variable cost financing."[31]

A closely related motive is to maintain capacity utilization.[32] The intention is not to capture market share but to respond to high world supply by accepting lower prices. The difference with dumping is that it is the buyer, facing excess supply and eager sellers, which sets the level of discount. Given the cyclicality of steel demand and the industry's rigidities, the phenomenon is not infrequent, and when it occurs, if the surplus is sufficiently large, the results can be dramatic. These results can occur even when steelmakers slash production in the face of collapsing demand, as the combination of production cuts and rapidly falling prices showed in the beginning months of the 2008 global financial crisis.

Dumping

International economics has produced a technical and sophisticated literature on dumping. The purpose here is to review the basic concepts just enough to understand the motives and consequences at issue. These provide a necessary background for examining the integrated steelmakers' policy preferences (taken up later in this chapter), steel as a globalized industry (the topic of Chapter 4), and anti-dumping actions as a practical protection (a topic of Chapter 5).

The standard typology of dumping was formulated in 1923 by economist Jacob Viner, and his central ideas on the topic remain a basic point of reference for current theory.[33] Viner's typology identifies both time frames and motives, making it a very useful introduction. Sporadic dumping arises when producers dispose of "causal overstock" – unplanned surpluses. To avoid depressing prices in the home market, the overstock is exported. Sporadic dumping can also occur unintentionally when prices in an export market rise without notice. Because the impact of these events is brief, neither need be regarded as injurious. Short-run or intermittent dumping occurs when producers respond to adverse changes in an export market by cutting their prices to keep market share. Alternatively, exporters seeking to become established in a new export market may discount to attract customers and establish goodwill. These two forms are more serious because their duration is long enough to affect domestic producers. More serious still is dumping intended to eliminate competitors in the export market or to impose a barrier to entry for new competitors.[34]

Long-run or continuous dumping occurs when producers wish to maintain full production without cutting prices in their home market. The situation arises when demand in the home market falls over time. Long-run or continuous dumping can also occur when producers seek to gain economies of scale but do not wish the extra production to oversupply the home market. Finally, state-sponsored dumping occurs when governments support discounted exports to improve trade balances or to help their firms expand abroad. Interactively, dumping can also occur as retaliation for dumping by others.[35]

According to Viner, only short-run dumping is a matter of concern because its duration is long enough to injure a domestic industry without necessarily supplanting it with a dependable and cheap foreign supply. Worse, since the duration of dumping is uncertain, producers, hoping to survive, may not redeploy their assets, thus imposing a welfare cost of wasted resources; still worse, they may carry on past the point of adjustment. Viner acknowledged that long-run dumping could indeed drive out domestic industries, but he also acknowledged that the change could shift production to more efficient producers and yield a net welfare benefit. In the same way, dumping to achieve efficiency through economies of scale would not be objectionable unless it resulted in the failure of otherwise efficient domestic producers. Long-term dumping of intermediate goods, such as steel, could even produce a net benefit if the result were dependably lower costs for domestic producers of final goods.[36]

The assumption is that consumers benefit from lower prices. If that benefit is provided only long enough to drive out competitors and hike prices, dumping is objectionable because the intention is to create a monopoly. Preventing monopolies, in Viner's view, is the clearest reason to oppose dumping. Recent economic scholarship agrees.[37] Indeed, anti-dumping rules are based on a long tradition of antitrust regulation and antedate the political consensus supporting liberalized trade.[38] At the same time, price discrimination between home and export markets may improve welfare by encouraging firms to produce more, making available goods that otherwise would not be. The same principle was just seen more generally as an offsetting benefit of price discrimination. If discounts and added sales were not available in export markets, producers enjoying high prices in protected home markets might find it preferable to sell only there. That benefit does not obtain as clearly when firms discount to maintain long-run full production without undercutting home prices. Because the result is both additional sales abroad and restricted home-market output, the global welfare effects are "ambiguous."[39]

Since dumping is price discrimination across national markets, a key consideration is whether the exporter's home market is open or closed to foreign competition. If it is closed, then charging different foreign and domestic prices is indeed possible. In the standard model, the home producer enjoys monopoly advantages in a protected market and export opportunities in open ones. Home markets may be closed by protective tariffs, and although these are falling as trade liberalizes under WTO agreements, they may still be significant in industrializing countries. More about these countries and the steel trade will be seen in Chapter 4. When export markets are open, the incentives to price discriminate may operate even when monopoly advantages and protections are weaker. Firms would still calculate prospective export revenues at the margin and compare them with the home-market alternative.

A key related condition is differing price elasticities in home and export markets. If prices in the export market are more elastic than in the home market, then foreign market sales will be more responsive to discounts, and prospective sales will be greater. Although demand for steel generally is inelastic, as was seen in the last chapter, the demand for any particular producer's offerings, as was also seen, is elastic because sales can be captured from others by cheaper prices. Revenue from export sales need not be as great as in a captive home market. Any revenue above marginal cost is profit. Again, these prospects are strongest when a producer has monopoly pricing power in its home market. Either a single monopoly firm or a disciplined oligopoly will be able to set production levels in order to extract prices well above marginal cost and earn higher profits than would prevail under competition. That surplus will enable them to sell abroad at enough of a discount to attract buyers and still cover costs. The losers are home-market consumers. If the discounted export is an intermediate good, home-market producers of final goods will be at a disadvantage if they export into competitive markets.[40]

Home markets may support higher prices for more innocently commercial reasons. Reputations are likely to be better established for home producers, who may enjoy the advantages of brand recognition, dependability as suppliers, and product quality.[41] Lower export prices may occur innocently as well. That can happen when conditions in the export market shift while the goods are already in shipment. If the goods cannot be stockpiled or if there is no feasible third market for them, they need to be sold at whatever price they can get.[42] Market uncertainty may also make it difficult to set accurate prices in advance. When firms must commit to particular levels of output before market conditions are sufficiently clear to allow proper estimation, they may end up with output that will sell cheaper than expected and still cover costs.[43] That price may undercut producers in the export market. The same result can occur when profit-maximizing managers set over-optimistic estimates and are forced to sell the excess product at a discount. If the producer balances sales of the surplus between a higher-priced home market and an export market, the latter price may be allowed to fall below marginal cost.[44] Both these examples arise from the more general task of setting price and output under uncertainty – a process that has fascinated economists.[45] As another way of dealing with uncertainty, an exporter may "self-insure" sales by setting a lower price for markets that are particularly volatile and risky.[46] In yet another form, uncertainty about respective market conditions and production costs may make it difficult for home producers to know whether an import is being discounted unfairly or simply reflects more efficient production. In Viner's view, only the former may be objectionable.[47]

Regular commercial activity in export markets may also produce price differences. Prices there may not be openly established but "negotiated and treated as trade secrets," resulting in different prices among otherwise identical products. Differences may also be a result of domination of distribution by particular firms and the presence of long-term contracts. When these practices produce artificially high prices for domestic products, normally priced imports may seem unfairly cheap. Moreover, secrecy in these practices may make it difficult for exporters to know what price to set. Large and powerful retailers, such as Wal-Mart and Home Depot, may generate price differences by forcing exporters to accept prices below their home-market levels.[48] Finally, normal exchange-rate fluctuations can raise export-market prices and make import prices appear artificially low.[49]

Dumping may also occur as a result of normal fluctuations in the business cycle. One imbalance arises when demand falls in the foreign producer's home market. The choice is between selling domestically for low prices or selling the surplus abroad. If a producer has decided to set production at a given level, has an inflexible production process, or must maintain particular output to cover costs, a handy expedient is to discount exports. That solution allows production levels to be maintained without lowering domestic prices.[50] These incentives are strongest in industries – steel, for one – with high fixed costs and cyclical demand.[51] A related incentive is to avoid disturbing labour relations.

The beneficiaries of cyclical dumping are the buyers in the export market, and the affected parties are the producers there. Even when these producers suffer losses as a result of discounted imports, the results are tolerable as long as the producers remain able to continue operating once the business cycle shifts to more favourable conditions.[52] The opposite imbalance may also occur: an export market is depressed while the home market remains sound. A producer that can cover costs domestically may be able to afford to abandon foreign sales. In that case, a cyclical fluctuation would lower imports in the foreign market.[53]

Strategies

Firms may discount in export markets for more strategic reasons. One is to exploit the excess profit margins of foreign oligopolies. If their profits are big enough, offering discounts may be quite worthwhile, even for firms with high fixed costs and small margins.[54] Other strategies are more forward-looking. Discounting, even to loss-making levels, can be viewed as a planned investment in future export-market share. This strategy is particularly attractive in markets that are growing rapidly.[55] Another future-directed strategy is to "forward price" exports below current costs in anticipation that costs will fall as scale economies rise.[56] Looking ahead as well, firms may discount their way into a market not only to sell their product but also to learn about the market itself and its prospects for future profitability. One motive is to identify weak competitors that are vulnerable to discounting.[57]

There are more directly combative possibilities. Utilizing home protection to ensure profits, a firm invests in technology and product development to gain advantages in cost and design. To achieve economies of scale, the firm raises output and sells the surplus abroad at a discount. The strategy is not only to gain market share and lower costs but also, by demonstrating scale economies, to discourage foreign competitors from risking the same kind of investment in their own markets. As was seen in the previous chapter, economies of scale can be a barrier to entry. This strategy is effective when the foreign producers decide that the exporting firm has compiled too many advantages. At its most successful, this strategy creates monopoly power in a foreign market.[58] In a variant of this strategy, an exporter discourages potential rivals by deliberately carrying excess capacity and showing an ability to charge low prices. In yet another variant, an exporter discounts to make rivals believe that home demand is low and that high export volume is necessary. The strategic calculation is that these rivals, expecting high volume, will cut back their own production.[59]

Governments may participate by giving subsidies to industries to pay uncovered costs. Expansively, subsidies may be paid to stimulate exports for the same strategic purposes just seen: to encourage development of economies of scale, to underwrite expansion into export markets, and to stimulate employment and regional development at home. Governments in industrializing countries may subsidize steel producers to escape dependence on steel imports. Less directly, the high minimum efficient scale of steel production may oversupply a domestic market and necessitate exporting the surplus. Subsidies to achieve scale economies may produce that result even when export development is not an objective.[60] A more clearly mercantilist motive for subsidies is to displace foreign producers in their own markets and enable home producers to capture the returns.[61]

Defensively, subsidies may be paid to keep uncompetitive industries from failing, either to cover production losses or, more optimistically, to underwrite modernization. In Canada, the former motive long supported subsidies to Sydney Steel Corporation in Nova Scotia, and the latter supported government aid for modernization at Algoma Steel in Ontario. A primary motive in defensive subsidization is to forestall plant closures, particularly in already depressed areas. This was true as well in Europe when its steel industry was undergoing restructuring and downsizing in the early 1980s, with the Belgian, British, French, and Italian governments paying some $35 billion to their steel producers while imposing quotas to manage supply and prices. Even then, as long as surplus capacity remains and subsidies offset costs, adjustment can be achieved by exporting at a discount.[62]

The cyclical nature of steel demand makes surplus capacity a regular prospect. Until the Chinese market began demanding large amounts of steel in 2004, which tightened supplies and raised prices, the world market was carrying an estimated 300 million tonnes annually of overcapacity. For industrial-country members of the OECD, the problem was severe enough that they agreed in December 2002 to stop subsidizing their steel industries except to fund worker retraining and environmental cleanups.[63] The problem addressed by subsidies in the face of excess capacity stems from the integrated steel industry's cost structure. As was seen in Chapter 2, the industry has a high minimum efficient scale, which requires high output. When excess output depresses prices, steelmakers are unable to cover costs, and when those situations occur, governments may intervene to support prices. In the 1980s and early 1990s depressed output and pricing were widespread. Steelmakers in the EU were operating at less than 75 percent capacity, and Japanese producers were doing only slightly better. Those conditions gave rise to both government intervention and export-price discounting.[64] By 2008, as was seen in Chapter 2, two of these conditions had moderated. Consolidation, improvements in efficiency, and strong profits in previous years enabled some steelmakers to operate at 65 percent of capacity – compared to 80 percent earlier – and still be profitable.[65] Global excess capacity was estimated to be 50 million tones.[66]

Nonetheless, for the affected producers, subsidized exports can be even more potent threats than merely discounted ones. That result was found empirically in a study that compared stock-market reactions to news of initiated trade complaint investigations involving subsidies and to news of investigations involving dumping. Subsidies were found to provoke a much sharper reaction: "When a foreign industry must contribute to the gaining of a competitive advantage [through dumping], the market views it as less foreboding than when the government undertakes the total expense as in a pure subsidy case." The reason is that governments have much greater resources than do individual producers.[67] Under WTO standards, permitted but actionable subsidies that cause harm to fellow members may justify the remedy of countervailing tariffs. Although steelmakers maintain that foreign subsidies are an issue, they have focused their efforts on dumping, which accounts for the great majority of their trade complaints. One reason is that showing harm from subsidies is more difficult than showing harm from dumping. Because subsidies are at least implicitly at issue, a brief look at recent economic treatments is useful. A survey of Canadian industry in 1995 showed the primary-metals sector, which includes the steel industry, to have been subsidized at a level below the business-sector average since 1968, varying between .1 and .4 of that average.[68] Such an industry would be particularly attentive to higher subsidies elsewhere.

The standard argument against subsidies is that, if there were investment gains to be had, financial markets would be willing to allocate resources. Unwilling markets mean either that the prospect is below investment level or that alternatives offer more returns or less risk. To have a net welfare benefit – the justification for spending public money – the profits made possible must exceed the amount of the subsidy. For the excess profits to be retained in the home country, the subsidized firms must also be domestically owned. Aside from the possibility that those profits may not occur, resulting in a net welfare loss, there are two important practical drawbacks. First, the necessary profit levels are possible only in an oligopoly, which must remain closed to new entrants attracted by the extra returns. If newcomers are allowed, they will reduce the profits through normal competition and dissipate the intended surplus, diminishing both their own returns and those of their rivals.[69] Second, subsidizing an industry early in its development may distort incentives by attracting too many firms and diverting resources from more naturally profitable and efficient applications.[70]

The beneficiaries of subsidized exports are consumers, for whom the underwritten discounts are a gift from the exporter's government. Receiving a bonus as well are producers in the export market for whom the subsidized product is an input.[71] At the same time, allowing foreign producers to "solve their excess capacity problem by selling cheaply," or allowing foreign governments to support aggressive expansion abroad, may create the same result seen earlier with short-run dumping: in the country receiving the subsidized imports, otherwise viable producers may be driven from the industry, production capacity may be unnecessarily reduced, and consumers may be harmed by the foreign producers' added market power.[72]

An important consideration is whether strategic dumping – with or without the sponsorship of foreign governments – warrants government intervention, and whether anti-dumping measures are a suitable response. One important issue is short-run dumping, which Viner believed to be the most frequent and harmful form. It lasts long enough to drive out domestic firms but not long enough to ensure a reliable and fairly priced foreign supply. Uncertain duration induces domestic firms to keep their productive factors employed and accept losses in the hope that the episode will be short enough for them to survive. That hope keeps them from seeking alternative uses for their assets or cutting losses and exiting the industry.[73] That conundrum, and the associated welfare costs of idled or underutilized capital resources, in Viner's view, may justify government intervention. Addressing that view are two contemporary economic arguments.

One argument against intervention emphasizes factor mobility. Modern factors of production are actually quite mobile and readily redeployed. They stay idle only if they have no better use. Even then, losses to owners of idle factors will be outweighed by the consumer benefit of cheaper imported goods. If firms cannot make a ready adjustment, then market imperfections must be present. These may be difficult for a government to identify, making it questionable whether appropriate countermeasures, including anti-dumping duties, will be applied.[74]

The opposing argument holds that easy factor mobility may exist in a hypothetical frictionless economy; in the real world, however, entry and exit barriers hinder redeployment, and factor prices may limit substitution at comparable rates of return. When that occurs, as Viner asserted, factors will remain underemployed or idle, and the economy will bear the welfare effects of that waste. This result is especially likely in industries, such as integrated steel, that have expensive plant, continuous production processes, large and highly specialized facilities, and high factor costs such as labour.[75] As well, short-term dumping may disrupt the price information that firms use for investment decisions. Useful opportunities may be forgone, and faulty allocations may be made. That result is also a welfare loss. Anti-dumping actions may not be the best remedy, but the real costs of short-run dumping should be acknowledged.[76]

These two views reflect a more general debate about free trade and protectionism. Supporting the free-trade position are prospective efficiency gains in production and resource allocation. These results, according to a large, cross-national study, do indeed prevail. The "turbulence" resulting from adjustment – increased market entry and exit, firm reorganizations and product and technology innovations – were positive and beneficial. Equally important was the discouragement of oligopoly.[77] Supporting the free-trade position as well are the benefits of scale economies. As prices fall with rising volume, commodities for export increase along with the ability of consumers to afford them. These benefits may accrue even if they are made possible by a protected home market.[78] At the same time, if such protection does not encourage domestic output, it will also not encourage exporting. That empirical result is strongest for industries with the highest returns to scale.[79] An important factor is producer complacency in a protected market.

When producers seek to displace rivals in foreign markets, their activity shades into predation. A standard definition of predation is "short-run market conduct which seeks to exclude rivals on a basis other than efficiency in order to protect or acquire market power."[80] This is done either by price discounting, which lowers rivals' income, or by other actions that raise the rivals' costs. More about those other actions will be seen presently. As in domestic markets, predation occurs when a firm sets export prices low enough, and persists with them long enough, to drive out competitors. The move depends on having enough staying power either to destroy the competitors' ability to continue or to intimidate them into withdrawal. One calculation is that the competitors' home financial market, in light of troubled prospects caused by the foreign producer, will be unwilling to cover losses.[81] Like predation among purely domestic producers, this behaviour is objectionable because it enables the victor to set monopoly prices. These have harmful welfare effects because monopolists can set higher prices than would prevail under competition by restricting output. Buyers must pay more, and supply is held below the level of actual demand. For that reason, preventing and punishing monopolistic practices has long been a staple of competition law. Harm may also occur below the level of predation. Even when discounting leaves surviving producers in the export market, the result may still be an increase in market power that supports higher prices and harms consumers.[82]

In OECD members' domestic competition laws, anti-predation rules "typically prohibit abuse of a dominant position, efforts to monopolize a market or price discrimination to injure competition in a market."[83] One problem in designing competition law is setting the proper threshold. Normal pricing behaviour should not be treated as predation. Laws that are too strict may not only discourage price competition but also actually promote collusion. Furthermore, they provide a way for firms to intimidate rivals with the prospect of investigation or prosecution. Economists term such behaviour non-price predation, and its purpose is to raise rivals' costs through tactics such as litigation. Predatory pricing, in contrast, is intended to lower rivals' income. A looser standard, on the other hand, may allow genuine injury.[84] These same possibilities are present in anti-dumping law, whose purpose is to curb predation by foreign competitors. An OECD report directly links the two: "increasing concern over international trading practices could benefit from clear thinking about what pricing should be considered predatory."[85]

In both competition and anti-dumping law, the key task is to determine whether predation is actually occurring. This entails both judging the actual merits of complaints and – because "rules against predatory pricing and dumping are the natural weapons of the non-price predator" – impeding the frivolous use of such complaints.[86] The fact that predation is intended to be hard to detect complicates the task. Two evidentiary requirements are specified in the WTO's anti-dumping standards: that the goods are being sold below home-market value or below the cost of production, and that they are causing injury to domestic producers. Those are the points that complainants seek to establish. Whether anti-dumping laws are necessary defences against unfair foreign competition or, rather, are easy ways of getting unmerited protection and engaging in non-price predation is a serious question for trade policy. Chapter 5 examines the debate. For now, what has been seen is that pricing behaviour in foreign markets may be far less straightforward than one might imagine.

Policy Preferences

Political economists are interested in the preferences of particular industries for open or protectionist trade policies. More will be seen in Chapter 5 about the Canadian and American governments' patterns of granting anti-dumping petitions. Of interest now is basic industry orientations in light of the pricing and trade conditions just seen. Affecting those orientations are the industrial and market characteristics outlined in Chapter 2.

Basic economic conditions frame the question generally. Over the long run, growing economies will support more open trade, and declining economies will support protection.[87] That proposition reverses a standard account in which tariffs set the level of trade. In this view, the level of trade, and trade prospects, set tariff levels. The key dynamic is long-term growth processes. Positive ones support long-term openness.[88] Basic political orientations also matter. One orientation is the acknowledgment of economic interdependence. In Rogowski's words, "in an age of objectively increasing dependence on trade, states will be increasingly constrained to remain open and to adopt the institutions that ... conduce to openness and effective competition in world markets." Governments require insulation from protectionist political pressure, independence from "rent-extracting firms, classes or sectors" that benefit from closed markets, and an ability to pursue consistent policy over time.[89] Evidence that these basic conditions are important has been seen in the resistance by governments of wealthy industrial countries to protectionist pressures during times of severe distress. A study of one such period, the 1970s, found much lower levels of protection than would have been expected.[90] One reason is a long-term commitment to open trade and to international agreements that support such trade. Another (just noted) is relative political independence from domestic protectionist pressures.

At the same time, governments must be attentive to serious dislocations caused by trade adjustment. The question is how to deal with failing sectors:

"Are old industries to be treated as social pensioners – like peasant systems of agriculture?" For governments facing structural transformation in the world economy, the task is to manage unavoidable economic change and to slow it down "whenever it exceeds a given pace." Adjustment is the most difficult and expensive in industries that have high fixed costs, low marginal costs, and large, immobile factors of production – conditions that amply characterize integrated steel. Expectations are also relevant. Short-term protection and financial assistance to modernize and adjust to current markets, when granted during times of distress, may carry "the implied promise that they will never be totally eliminated."[91]

Trade policy preferences do vary by industry. A strong influence is a firm's position in the international economy. Favouring open trade are firms that export and import materials, components, and finished products, that have internationalized production systems, and that trade internationally with other branches of the firm. Additional reasons to prefer open trade include these: a vulnerability to retaliation; a concern that protectionism in other countries will divert their exports to less favourable third markets; an appreciation of the extra transaction costs that protectionism would impose; a desire to keep tariffs on imported supplies and components low; and a concern that "domestically oriented" national competitors would gain unduly by protection. The converse also holds: domestically oriented firms, particularly those under pressure from imports, will prefer protection as a less costly strategy than adjustment and modernization.[92]

A basic economic question is whether factors of production are mobile or immobile. Mobile factors of production may be redeployed in the face of import competition, while immobile ones face loss-making operations or abandonment. As was just seen, that consideration was central to Viner's views about dumping. Specialization may increase factor immobility both of specialized labour and of specialized firms themselves – neither of which are "notably adaptable."[93] In the same way, capital intensity imposes high entry and exit barriers, limits adjustment, and carries high fixed costs.[94] These conditions all impose high costs of adjustment. Limited factor mobility and imperfect markets may encourage popular support of trade policy as a "kind of social insurance, aimed at insulating the ... economy from injury caused by adjustment stimuli abroad." A related encouragement is a low dependence on export markets.[95] These points derive from the specific-factors – or Ricardo-Viner – theory of trade. That theory posits that factors which are specialized for producing particular commodities and which are not readily transferred to other uses will bear trade's full impact, both positive and negative.[96] Policy preferences derive directly from the level of impact and from the degree to which these assets are competitive on world markets. The more these assets are "stuck in their present occupations" and are threatened by trade, the greater is the incentive for their owners – both capital and labour – to press for protection.[97]

Also important is whether an industry produces intermediate or final goods. A survey of world anti-dumping cases between 1987 and 1997 found that almost 70 percent involved four industries: base metals, chemicals, machinery and electrical equipment, and plastics. In the base-metals cases, the majority involved steel. One explanation is vulnerability to cyclical price changes, leading firms to price exports below costs during the bottom of the cycle.[98] As was seen, the steel industry's economic characteristics strongly dispose it to that behaviour. Also pertinent is resource content. Agricultural products, steel and other metals, chemicals, and forest products, which have a high resource content, are "reflected disproportionately" in anti-dumping and unfair-subsidy cases.[99] One interpretation is that these commodities have both international and domestic suppliers and have high levels of substitutability.[100] Heavy trading of these goods accordingly generates complaints.[101] A constraint occurs, according to economist Bela Belassa, when substitutability supports intra-industry trade – a two-way exchange of highly similar or standardized goods. The United States and Canada conduct an intra-industry trade in steel. Although Belassa asserts that governments "cannot export and protect the same commodity,"[102] both the United States and Canada have recourse to anti-dumping remedies under the North American Free Trade Agreement (NAFTA) and have used them against each other's steel exports.

Producers of final goods have opposed interests. When import barriers are raised for an intermediate good, its price rises. That enables domestic producers of the same good to raise their prices as well, and producers of final goods face higher materials costs. If they can pass along the increase to consumers, they will have little incentive to oppose protection. That changes if they must absorb the costs, particularly if they compete on international markets against lower-priced products.[103] Important as well is whether the dumped intermediate good is homogeneous and highly substitutable.[104]

Another structural consideration is an industry's level of concentration. As was seen earlier, industries that are highly concentrated as monopolies or oligopolies enjoy pricing power in their home markets. These industries prefer protection because a competing import, particularly one offered at discount, forces prices down and reduces pricing power and profits. The same incentives can arise in industries with low levels of concentration if they are enjoying high home-market demand and are operating at high capacity utilization. For them the incentive is to capture the returns from demand and scarcity by restricting imports.[105]

Vulnerability to large swings in the business cycle affects preferences. As was seen, business cycles can generate dumping as foreign firms seek to cover costs by selling abroad at a discount. The effect on domestic producers multiplies if their market happens to be in a cyclical trough at the same time. Industries exposed to such swings will prefer policies that allow the government to impose anti-dumping remedies during adverse swings in the business cycle. Particularly interested are industries, such as steel, that have expensive and long-run capital investments to cover.[106]

For the integrated steel industry, several of these conditions stand out. Steelmakers have highly specialized factors of production, and these reduce their ability to adjust to import competition by shifting to products with stronger advantages. Reducing capacity, another adjustment, is limited by high exit barriers and ongoing fixed costs. Modernization may be beyond the resources of weakened firms. These conditions limit adjustment to trade and support preferences for protection.

At the same time, other conditions associated with protectionist preferences do not apply. Labour-intensive producers that are unable to compete with high labour-content imports adjust by becoming capital-intensive. Inability to do so supports preferences for protection. Steelmakers did indeed make that adjustment and are now highly automated. Outsourcing uncompetitively costly parts of production is another adjustment. That option is not readily available to steel because of high production indivisibilities, although some mills, including integrated ones, have begun importing slabs – an intermediate steel product – to finish into higher-end and more profitable final products. As will be seen in the next chapter, doing so outsources that part of the process to foreign mills. Altogether, while some of integrated steel's basic structural characteristics align with protectionist preferences, not all do. That mixed picture is worth keeping mind when Chapter 5 reviews anti-dumping as a survival strategy.

In an effort to connect preferences to policy outcomes, some international political economists have focused on the interplay between the interests of societal groups and elected governments, and have employed some basic reasoning from microeconomics. Interests are defined as utility functions, and actors are defined as utility maximizers. Groups of producers are assumed to be unitary and rational in their behaviour, as are national executives and legislatures, the political actors.[107] For producer groups, utility is defined as net income, which is revenue minus the cost of the goods they consume and the cost of the production inputs they use. The desire to maximize net income generates policy preferences accordingly. Groups will support policies that increase income and lower the costs of inputs and consumption goods, and will oppose policies that reduce income and raise those costs.[108] Utility for national executives and legislatures is defined as political support and re-election. These actors will prefer policies that provide such utility and oppose policies that do not. Important is the condition of the overall economy and of the groups on whom the political actors depend for support.[109] In a "process of market exchange," responsive attention is accorded to groups that are the strongest and best organized.[110] Policy outcomes result from the utility-based strategic interactions among producer groups, political executives, and legislatures within the prevailing institutional structure.[111] Outcomes reflect a relative balance of power among organized interests. Among groups advocating open trade and those advocating protection, the balance shifts according to the business cycle and the overall performance of the national economy.[112]

"General structural theories," observe Katzenstein, Keohane, and Krasner in their evaluation of international political economy, "encounter numerous empirical anomalies."[113] Rational-utilitarian theory, as one kind of structural theory, may not account for the ability of governments to act according to a more general public interest and to take a long-term view.[114] Indeed, through a commitment to open trade and a reluctance to grant more than marginal concessions in the face of protectionist pressure, governments have produced an empirical record of "too little and too late." Such a record would not be predicted by the rational-utilitarian theory just seen. One reason is that key actors in trade decisions are not political executives and legislatures but trade bureaucracies.[115]

In both Canada and the United States, trade bureaucracies are indeed the key actors in anti-dumping proceedings. As will be seen in Chapter 5, investigations of industry complaints and assessments of injury are conducted in administrative tribunals. Although there is debate over how susceptible these tribunals are to influence by interest groups and elected governments, it can be said that they do their work according to a set of rules and procedures. The formal adjudication of disputes, including disputes over dumping, serves to insulate them from the pressures and short time horizons of electoral politics. Anti-dumping rules and procedures have been criticized as facilitating positive findings and thus as favouring complainants, but that, too, is debatable, as will also be seen.

More generally, structural approaches have been criticized as "static and ahistorical."[116] One reason why steel producers, along with other industries affected by import competition, have opted for anti-dumping proceedings instead of lobbying is that the costs are lower and the prospects of success higher. The fact that such choices exist makes trade politics more potentially variable over time than strictly rational-utilitarian models might lead one to imagine.[117] Producers, also acting on rational-utilitarian grounds, may decide that investing and modernizing is a better strategy than protection, or they may decide to pursue both strategies. Modelling the institutional complexity that allows such latitude is a daunting task.[118] At the same time, basic industrial and market conditions do matter. Practical realities of steel production, and the pricing and trade behaviour just seen, establish conditions that no sensible management can ignore. In that way they are important to know.

Conclusion

Competition in any industry carries risks, but basic conditions in the steel industry add extra dimensions. Because steel production has high fixed costs and low variable and marginal costs – a key condition explained in Chapter 2 – producers facing oversupplied markets adjust not by cutting production but by cutting prices, although in the short run, under conditions also seen in Chapter 2, they may be able to cut both. Because all must cover their fixed costs, their ability to sustain periods of low prices is limited. That vulnerability can be exploited when producers enjoy assured profits in some of their markets, enabling them to cut prices in others. Such price discrimination is possible internationally in any industry that enjoys home-market stability. Incentives to price-discriminate in export markets are numerous both theoretically and in practical terms. The steel industry's cost structure and its limited factor mobility make it vulnerable to discounting, including that by foreign exporters. One result is the relatively frequent incidence internationally of steel in anti-dumping proceedings.

The necessary scale of steel production concentrates the industry among a few large producers. When they are able to co-operate according to common incentives and understandings, they can maintain profits and cushion downturns. The arrangement ends when producers that have no incentive to co-operate enter the market. A normal way for newcomers to claim market share is through attractive pricing, particularly when the product at hand is highly substitutable. When price discounting is predatory, both domestic and international trade laws forbid it.

Dumping has both unintentional and deliberate origins. Prices below home-market level or below the cost of production may occur as consequences of market changes, or they may be set to gain market share and to intimidate foreign competitors. Dumping is a matter of concern for governments if it results in monopoly power. Preventing monopolies is the purpose of domestic competition law, which is anti-dumping law's immediate ancestor. As a prospective form of protection against imports, anti-dumping law also aligns with trade-policy preferences. Several important conditions in the steel industry do support preferences for protection, but not all. One exception has been the industry's willingness to adjust to foreign competition by improving labour productivity – even though the industry's capital structure makes modernization quite costly, as Chapter 2 explained. Another exception instead, is conceding uncompetitive product lines. These warrant caution in using structural trade theories. At the same time, the steel industry has been one of the top generators of trade complaints internationally, and this makes pricing, preferences, and government action pertinent. This is so because steel has become an internationalized industry producing an important and readily traded good.

These matters converge on the issue of viability. For individual firms, viability means being able to survive unfavourable pricing cycles, which discounted imports may exacerbate. For national economies, viability means preserving competition. Discounted imports may destroy otherwise sustainable firms and industries, but they may also limit the pricing power of natural oligopolies. When trade and pricing produce conflict, the interests to weigh are those of the producer and those of the public. A sophisticated economic literature, reviewed in Chapter 5, evaluates that question. The behaviour giving rise to it has been seen here. Next to examine is the diffusion of the steel industry from Europe and North America to Asia and Latin America and the rise of steel as a widely traded commodity. Chapter 4 takes up that development.

4
Trading Steel

The equipment for making steel is portable, efficient, and for sale. That has made it possible for countries beyond the steel industry's historic homeland in Europe and North America to develop industries of their own and install the latest technology. Although the motive was often to develop domestic steel suppliers for emerging industries, exporting became an option as the steelmakers became established and began to expand. In two instances, postwar Japan and later South Korea, exporting was a key requisite for national economic growth. Other producers have found exporting to be a useful adjunct to home markets. As a highly standardized product, commodity-grade steel can be sold globally, and sixty years of trade liberalization have opened markets abroad.

The traditional theory of comparative advantage explains trade as the result of favourable and complementary factor endowments among trading partners, in which one's efficiently produced good is exchanged for the other's efficiently produced good and in which partners concentrate their resources where their respective efficiencies are greatest. Such endowments need not be purely natural. States, through sponsorship of new industries, have the ability to secure and reinforce conditions that may not produce trading advantages on their own. A newer theory based in economic geography acknowledges that factor and explains trade in terms of economies of scale and transportation costs. When these are favourable they allow goods to be produced and shipped cheaply, making them competitive in external markets. Transportation costs, represented as forward and backward linkages, are used more generally to account for the rise of industrial regions and for proximate and remote sources of supply. This explanation, because it takes account of state involvement and the role of distance, applies strikingly to the steel industry, as will be seen. Chapter 2 showed how the industry's technology and organization encourage economies of scale.

In 2007, 36 percent of the world's production of finished and semi-finished steel was traded, amounting to 435.5 million tonnes.[1] High volumes are nothing new. Steel imports have "long formed a substantial portion of the total supplies of steel coming into the market in nearly all the major consuming countries. They are part of a settled pattern ... not an aberration."[2] In 1975, world steel production was less than half that of 2007's, and of that, 22.6 percent was exported.[3] Demand for imported steel arises when domestic production is inadequate, when purchasers desire a larger set of suppliers, when foreign sources are located closer to particular markets than domestic producers, and when import prices are cheaper.[4]

The amount of steel available to export is set to rise as industrializing countries add capacity to meet growing domestic demand. Using current high-output technology and exploiting available economies of scale, they will begin producing surpluses as domestic demand matures. China, by far the largest and fastest-growing producer, moved from being the leading net importer in 2003 to the leading net exporter in 2006. (Net exports are exports minus imports, and net imports are imports minus exports.) Behind China, the top net exporters are Japan, Ukraine, Russia, and Brazil, which together account for 81 percent of total net exports.[5]

Canada and the United States conduct an active two-way trade as primary suppliers and customers. Why that special relationship should prevail amidst an international population of steel exporters is explained by the close linkages that join producers of intermediate and final goods. Particularly important are the linkages between North American steelmakers and their biggest customers, the automakers. Also important is proximity, represented both as geographic distance and business familiarity. Border effects – the factors that encourage or inhibit trade – are considered in Chapter 6.

This chapter surveys current patterns of steel production and trade. To account for the industry's location in existing centres and for its recent expansion into industrializing countries, it will examine economic geography's account of industrial location, linkages, transportation, and trade advantages. The emergence of the Japanese, South Koreans, Brazilians, Russians, and Chinese as major steel exporters will be treated as a case in point. A review of cost differences between producers in developed and industrializing countries will follow. Those differences are less dramatic than one might assume, and standardized technology is the equalizer. If there are locational advantages in the steel industry, they relate to domestic iron-ore supplies. How these are distributed confers benefits on both sides of the north–south divide. The conclusion of this chapter will evaluate these facts in light of pertinent tenets of international political economy.

The Basis of Trade

In the industrial economies, use of steel in the automotive, appliance, construction, electronics, and machinery industries reflects advanced

Table 4.1

Top industrial-country net importers of steel 2006

Country	Net imports, mmt	Rank in top 15
United States	32.6	1
Spain	7.4	3
Italy	6.9	4
EU (25)	5.1	8
Canada	4.9	9
South Korea	4.4	10

Source: IISI, *World Steel in Figures 2008;* Major Importers and Exporters of Steel 2006, 18.

development.[6] The largest consumer of steel in 2007 was China at 408.3 mmt. It was followed by the EU at 193.2 mmt, the United States at 108.2 mmt, Japan at 80.1 mmt, and South Korea at 54.8 mmt. Canada's steel consumption, at 15.4 mmt, places it in the middle ranks with France, Mexico, Brazil, Taiwan, and the United Kingdom.[7] As an index of rapid industrial growth, India's steel consumption almost doubled between 2001 and 2007, with that year's amount, at 50.8 mmt, being just below South Korea's.[8] Net imports reflect those levels of consumption in light of domestic output. Table 4.1 shows the industrial countries among the top fifteen net steel importers.

Steel consumption in the industrial economies is variable, as Table 4.2 shows. Several developments account for this pattern. Russia's large increase reflects economic recovery and investment in the oil industry. These, plus exporting, have made Russia the fourth-largest steel producer.[9] The EU's and

Table 4.2

Steel consumption in the industrial economies, 1998-2006 (mmt)

Region	1998	2007	Increase (%)
EU*	151.8	193.2	27.3
United States	115.7	108.2	-6.5
Japan	70.3	80.1	13.9
Russia, Ukraine, and other CIS	25.7	51.9	101.9
Taiwan	20.2	18.1	-10.4
Canada	15.8	15.4	-2.5

* EU figures for 1998 list 25 members and for 2007 list 27 members.
Source: IISI, *World Steel in Figures 2005,* Apparent Steel Use 1998 to 2004, 20; *World Steel in Figures 2008,* Apparent Steel Use 2001 to 2007, 12.

Table 4.3

Steel consumption in the developing and industrializing economies 1998-2007 (mmt)

Region	1998	2006	Increase (%)
Asia	275.9	664.5	140.8
Central and South America	27.7	41.6	50.2
Middle East	15.8	47.6	201.3
Africa	15.7	25.4	61.8

Source: IISI, *World Steel in Figures 2005,* Apparent Steel Use 1998 to 2004, 20; *World Steel in Figures 2008,* Apparent Steel Use, 2001 to 2007, 12.

Japan's increases reflect the consumption and export of manufactured goods. These increases have occurred despite the automobile industry's use of thinner steels as it has moved to produce more fuel-efficient vehicles (the figures above measure weight) and despite the rising efficiency of continuous-casting technologies, which have increased the portion of raw steel transformed into finished products and reduced wasted metal by 15 percent.[10] At the same time, the steel content of manufactured imports reduces the use of domestic steel. Between 2000 and 2003, American steel consumption declined from 114.7 to 100.9 mmt, largely as a result of the movement abroad of metal fabrication, of which the automotive-components sector represents a significant portion.[11] These patterns can also change quickly. American steel consumption rebounded to 120.3 mmt in 2006, reflecting a more vigorous economy, but in 2007, reflecting a slowing economy, it fell to 108.2 mmt. Limiting further decline was a low US dollar, which stimulated exports of manufactured goods with a large steel content. By contrast, Canada's consumption from 1998 to 2007 was quite steady. A modest decline between 2006 and 2007 reflected an increase in manufactured imports and a decrease in manufactured exports, partly the result of a higher Canadian dollar.[12]

Among the developing and industrializing regions, consumption is rising briskly, as Table 4.3 shows.

In these regions, the sources of steel demand are at the high *and* low ends of the market. At the low end, the demand for long-steel products reflects the rapid construction of buildings and infrastructures and rising "steel intensity" – partly the result of upgraded building standards. At the high end, the automotive industry's growth is increasing demand for specialty and sheet steels.[13] Rising levels reflect economic growth more generally, and one index is the relationship between gross domestic product and per capita steel consumption. In 2001, for example, China consumed only 92 kg per capita, in contrast to Malaysia, which consumed 450 kg. Given the components of steel demand, GDP growth can be expected to have a very strong

Table 4.4

Steel consumption per capita in the industrial countries 2007 (kg/year)	
South Korea	1,135.5
Taiwan	781.8
Japan	625.9
Canada	469.3
EU	392.0
United States	353.9
Russia	279.9
Ukraine	172.7

Source: IISI, *World Steel in Figures 2008,* Apparent Steel Use per Capita 2001 to 2007, 13.

effect.[14] Results have borne that out. By 2007, China's per capita steel consumption had risen to 307.3 kg.[15] Table 4.4 compares that amount with those of other industrial countries.

South Korea's high per capita figure can be accounted for partly by the prominence of shipbuilding, whose major steel input is heavy plate. Regarding *overall* steel consumption, China's increased by 158.4 percent in the same period – 2001 to 2007 – accounting for 80.3 percent of Asia's growth. China consumes twice as much steel as the EU and nearly four times more than the United States.

China's rapid economic growth has generated excess steel demand. Its net imports peaked at 34.7 mmt in 2003. Its steel industry's equally rapid expansion reduced net imports to 13.1 mmt in 2004, and they fell to only 100,000 tonnes in 2005. In 2006, China became a net exporter at 32.6 mmt.[16] China's main customers are South Korea, Taiwan, Thailand, and Vietnam. Its exports to the EU rose rapidly from between 0.3 and 0.4 mmt in 2005 to a stunning 5.0 mmt in 2006, prompting vigorous complaints from the EU's industry commissioner.[17] In 2007, Eurofer, the steel producers' association, threatened to file anti-dumping complaints against Chinese steel imports.[18] China's exports to the United States jumped from 2.4 mmt in 2004 to 4.8 mmt in 2006. American producers, having drastically retrenched earlier, were unable to meet domestic demand, leaving market space for imports amounting to 30 percent of domestic consumption and making the country an attractive site for foreign steel producers to set up operations, as will be seen later.[19] Among the other industrializing countries, only South Africa, Kazakhstan, and India are in the top fifteen net steel exporters. Others are large net importers, as Table 4.5 shows.

Thailand's figure reflects export-based manufacturing, and Iran's reflects inadequate domestic production despite its 2007 ranking as the twentieth-largest producer. The UAE's figure reflects low domestic production and high

Table 4.5

Top industrializing net importers of steel 2006

Country	Net imports (mmt)	Rank in top 15
Thailand	8.5	2
United Arab Emirates	6.7	5
Iran	5.6	6
Vietnam	4.0	11
Saudi Arabia	3.7	12
Mexico	3.3	14
Turkey	3.1	15

Source: IISI, *World Steel in Figures 2008,* The Major Importers and Exporters of Steel, 2006, 18.

levels of construction. To put these figures into perspective, recall that Canada's net imports in 2006 were 4.9 mmt.[20]

Steel production is growing worldwide. In 2004, it exceeded 1 billion mmt for the first time, and in 2007 it was 1.34 billion tonnes.[21] With 2007 production of 489.2 mmt – a 120 percent increase over 2003 – China is by far the world's largest producer, accounting for one-third of total volume. China's output is four times that of Japan, the second-largest producer.[22] After China and Japan, the remaining top ten steel-producing countries in 2007 were, in order, the United States, Russia, India, South Korea, Germany, Ukraine, Brazil, and Italy. Canada, in sixteenth place, ranks between Mexico and the United Kingdom.[23]

As Table 4.6 shows, world steel production yields a mixed pattern of domestic shortfalls and exportable surpluses.

The EU exports steel, but it also imports lower-margin products and semi-finished steel. In 2006, its main suppliers were Russia, Turkey, Ukraine, China, India, and Brazil.[24] The top suppliers of steel to the United States in 2007 were Canada, China, Brazil, Mexico, and Japan.[25] Canada's top suppliers were the United States, China, South Korea, and Germany.[26] Rankings vary according to commodity. Higher up the product line, for example, the top five suppliers of alloy flat-rolled steel to the United States in 2007 were China, Germany, Sweden, Mexico, and Canada.

Overall, the industrial countries' steel imports reflect abandonment of lower-end products, where steel producers in the mature industrial countries have continued to shed capacity. Not all of the reduction has been voluntary. The American steel industry's thirty-three bankruptcies between 1997 and 2002 involved 46 mmt of capacity, of which some has been salvaged and recombined in new firms. For both the EU and the United States, the largest import categories are semi-finished steel and hot-rolled coil for processing

Table 4.6

Steel surplus and shortfall as percentages of world production 2007

Net Exporters	Surplus (%)	Net Importers	Shortfall (%)
CIS states	4.7	Asian countries (not including	
China	2.6	China and Japan)	3.8
Japan	2.3	NAFTA	1.9
Central and South America	0.3	EU (27)	0.4

Source: IISI, *World Steel in Figures 2008,* Steel Production and Use: Geographical Distribution, 2007, 15.

into higher-end products. In 2006, hot-rolled sheets and coils led the list of internationally traded steel commodities at 62.9 mmt, followed closely by ingots and semi-finished steel at 62.5 mmt, representing together over one-third of total volume.[27] That demand generates trade, and for producers of these commodities these figures represent sizable export markets.

Canada and the United States conduct a two-way trade in both low- and high-end products. Between January and May 2008, Canada imported $758 million of flat-rolled steel products that were not clad, plated, or coated, and it exported $482 million of the same. At the higher end, Canada imported $482 million of alloy-steel flat-rolled products and exported $270 million of the same. Overall, in 2007 Canada imported $7.2 billion of iron and steel from the United States and exported $5.8 billion.[28] Reflecting industrial integration between the two countries, steel producers buy from many of the same suppliers and sell to many of the same customers.

These developments raise basic questions of organization and locale: Why do industries arise in particular places? What is the role of distance? How does interdependence develop among the producers themselves? Recent economic geography has been occupied with these questions and their relation to trade theory. The main points shed useful light on world steel.

The Importance of Place

Traditional models of trade have been based on comparative advantage and constant returns to scale. Comparative advantage results from favourable combinations of factors. When these are concentrated in a particular locale, it is a naturally more efficient producer of the commodity in question than are producers in locales where concentrations are less optimal. Trade arises from the exchange of commodities among optimally efficient places of production. Constant returns to scale mean that no cost advantages accrue with increased output. Instead, costs are assumed not to decline with volume. That view fits the standard economic model of perfect competition in which no producer is large enough to achieve economies of scale. These two ideas formed the core of traditional trade theory, partly because the conditions could be modelled mathematically.[29]

Modern industrial specialization, however, is based on economies of scale. These, in turn, encourage large firms, and their workings in the integrated steel industry were seen in Chapter 2. Economies of scale also encourage imperfect competition, and Chapter 3 examined the effects of this on pricing. Adding the two conditions together yields a picture of specialized firms producing in high volumes. Production concentrates in particular locales not because of naturally optimal factor combinations but because of the advantages provided by specialization. Trade arises from increasing volume and scale economies. Low transportation costs enable these to be exploited. This pair of causal factors has been termed the economic geography model of trade.[30] Formulating that model won Paul Krugman the 2008 Nobel Prize in economics.

The conditions encouraging scale economies may be arbitrary and cumulative, so that a firm successfully produces, specializes, and gains market share. The growth of volume generates economies of scale. Increased pricing advantages support further expansion, both of volume and market domains, and provide revenue to improve technology. The result, in contrast to comparative advantage's natural fabric of complementary resource and factor endowments, may be artificial and occur when a successful firm locks in advantages that were originally unspecific and transitory.[31] Governments may intervene to assist that process.[32]

The same conditions encourage the growth of specialized industrial regions. A successful and expanding producer attracts related ancillary activities. As these activities develop and become efficient, the benefits of specialization and scale begin to accrue. The eventual result is external economies of scale, which are collective efficiencies that arise from proximate and connected activities. These in turn foster further efficiencies in the core producers and attract still more ancillary industries. Together with the advantages of proximate locations and low transportation and inventory costs, these conditions "simultaneously encourage spatial clustering and vertical integration."[33] When particular kinds of production cluster in a particular region, they attract population, create expanding markets, and induce more economic activity to locate there.[34] In this way, industrial concentration can be said to be path-dependent, in the sense that a particular cluster of activities, once established, sets the conditions that guide the region's development.[35]

Intermediate goods, such as steel, benefit from economies of scale. If those economies did not exist, production could disperse to smaller centres with no loss of efficiency. "It is only the presence of increasing returns that makes a large center of production able to have more efficient and more diverse suppliers than a small one."[36] Forward linkages connect intermediate-goods producers with producers of final goods, and backward linkages connect them with suppliers of materials and resources.

By locating close to their customers, intermediate-goods producers create short and direct forward linkages. And by locating together, intermediate-goods industries generate external economies of scale that benefit both them and their forward-linked customers. A set of forward and backward linkages encourages a large cluster of industry. Together, these factors can explain industrial regions as the products of complementary decisions among forward- and backward-linked producers.[37] Steelmakers, for their part, have forward linkages with manufacturers, particularly those of large durable goods, and backward linkages with ore and energy suppliers. Dense forward linkages encourage location near final-goods producers.[38] That proximity, as was seen in Chapter 2, accounts for regions of heavy industry such as the one bordering the Great Lakes.

Those same external economies may induce intermediate-goods producers to follow large shifts in demand. That explanation has been advanced to account for American steel production's historic expansion from its initial centre in Pittsburgh/Youngstown to Chicago as large new mills were built around the southern shore of Lake Michigan. Forward linkages to new manufacturing account for the shift.[39] More recently, the same linkages have shaped decisions on mill closures and rationalization. The largest closures have been in the Pittsburgh area, leaving behind a much smaller set of specialty-steel producers.[40] The relatively lower rate of mill closures in the Midwest is due to the concentration there of the industry's primary automotive customers. Canada's steel producers in southern Ontario have a strong export advantage because American customers in the Great Lakes industrial region can be supplied on a just-in-time basis.[41]

Forward linkages also account for the pattern of Japanese joint ventures with Midwestern steel producers in the 1980s. An important purpose of those ventures was to supply automobile manufacturers, who require close informational and material linkages with their steel suppliers. Most automotive sheet steel is made in Indiana, Michigan, and Ohio. Those same close linkages induced Japanese automakers to locate their own new facilities in the Midwest and southern Ontario. Their moves were part of a broader global strategy, derived from the *keiretsu* groupings of producers in Japan, of assembling regional clusters of manufacturing and supply in North America, Europe, and East Asia.[42]

Locale and Transportation

Low transportation costs encourage two patterns of domestic and international industrial development. In the first, firms exploit economies of scale by concentrating in centralized industrial regions and supplying dispersed markets from there. When those industrial regions are also population centres, large local markets may be served directly. Inexpensive transportation connects more distant markets.[43] That combination sustains existing clusters of production and centre-periphery patterns both within individual states and in global markets generally.[44] In the second pattern, low transportation costs allow firms to locate in peripheral regions where production costs are lower and to ship to major markets. The locational advantages are enhanced by economies of scale.[45] That combination fosters new industrial clusters and enables them to serve distant markets, creating multiple centres of production, equalizing income distribution, and reducing regional inequalities.[46] The same conditions operate internationally when producers utilize low transportation costs to serve export markets. One incentive to export is to develop economies of scale.[47] Interestingly, a recent econometric study found that economies of scale and transportation costs do not have "robust" effects in international trade, while those effects are very much present in interregional domestic trade. The two cases studied were trade among OECD members and between regions in Japan. The authors concluded that the difference is due to international trade barriers, which significantly limit the benefits of economies of scale and transportation.[48] Chapter 6 considers trade barriers and the future of the steel industry. At the same time it is worth remembering that 36 percent of current world steel production is exported.

Two sets of transportation costs affect the location of steel production: overland costs, which are relatively high, and overseas costs, which are relatively low.[49] High overland costs, together with external economies and short forward linkages, encourage integrated steelmakers to locate in major industrial regions. Once there, strong forces compel them to stay. Heavy sunk costs in existing facilities and the large capital costs of opening new ones in emerging regional markets have led the American and Canadian integrated producers to concentrate their investment in existing centres.[50] Likewise, the Japanese joint steel ventures in the Midwest were all with existing operations. Those same costs encourage minimills to move afield. Smaller capitalization makes new facilities easier to finance than new integrated mills, and serving regional markets is possible because minimills operate efficiently at lower volumes and require fewer linkages and ancillary services.

High overland costs make distant domestic markets expensive to serve. In the United States, a steel deficit appeared on the west coast because existing steel capacity was devoted to products for which there was no longer regional demand. Plans for integrated mills that would meet changed demand were abandoned because new facilities were too expensive. An attractive import market was the result. Although rail deregulation had reduced overland costs from the Great Lakes, Japanese and South Korean steelmakers shipped from mills at seaport locations, enjoyed comparatively favourable ocean freight rates, and landed their goods at the west coast's major markets.[51] Ample demand was provided by a strong regional economy.[52] In Canada, transportation costs to the west coast made the Great Lakes industrial region the integrated producers' market.[53] Minimill location reflects the same regional demand and the sector's locational adaptability. The former Ipsco Steel, which specializes in pipe products for the petroleum industry, has a large plant in Regina, Saskatchewan, and smaller ones in Calgary and Red Deer, Alberta. Ipsco was purchased in 2007 by Svenskt Stal AB, which sold the Alberta and Saskatchewan facilities to the Russian steelmaker Evraz a year later.

As American steelmakers began closing outdated and inefficient facilities in the 1980s, those nearest to seaports – even ones making higher-end products – were regarded as the most vulnerable to imports, and by the mid-1990s integrated steel had retracted into an inland industry with short and numerous forward linkages. The same regional proximity creates a single market area for integrated producers in the United States and Canada, a relationship demonstrated by the fact that the two countries are each other's primary source of steel imports.[54] More removed regional markets attract supplies – and producers – from abroad. These can follow industrial migration, and the American automobile industry shows how locational patterns can change.

Led by Japanese automakers, production facilities in the United States have moved to the southern states and raised that region's demand for automotive sheet steel to 45 percent of the industry's total for the entire country. In neither that region nor Mexico were there producers of exposed-sheet quality steel.[55] In 2004, Arcelor and Dofasco announced that they were seeking greenfield sites in Kentucky, Tennessee, and Alabama for a joint sheet-steel venture. Since Arcelor's steel would be imported from its European and Brazilian mills, seaport proximity favoured Alabama.[56] The plan was later abandoned after the two companies failed to reach terms, but a new joint venture is filling the demand locally. Russia's OEO Severstal and a group of American investors built SeverCorr, a minimill in Columbus, Mississippi, to provide sheet steel. The plant uses an electric furnace that supplies two ladle furnaces as well as a two-tank vacuum degasser – a combination that can produce automotive-quality metal. It also has a sheet-rolling line.[57] This is the first electric-furnace–based mill to produce automotive sheet steel. In yet another regional variation, Hyundai Motors supplies steel to its new assembly plant in Alabama from its mill in South Korea. Production requires 400,000 tonnes of finished steel annually, amounting to a shipping bill of about $20 million – a figure that does not include customs charges and delivery from port to the plant in Montgomery.[58] Finally, in 2007, Thyssen-Krupp announced that it would be building a new $3.5 billion facility in Alabama. To be noted in these cases are the locational elements: (1) a large regional demand for a specialized product, and (2) a balance between the costs of overland and overseas transportation and between foreign and local production. Given the automobile industry's relentless need to cut costs, all three cases represent efficient solutions. They show that costs may resolve in different combinations with divergent locational outcomes.

The bulk ocean-shipping industry has been characterized by the OECD as "relatively free and competitive." The industry comprises many small firms that operate as individuals; its members rarely form alliances.[59] A long-term and general decline in ocean bulk-transportation costs has made remote iron-ore and coal sources profitable and encouraged long forward linkages to steelmakers. By 1990, 90 percent of ore exports were seaborne (an almost 40 percent increase over 1983), with two-thirds of the tonnage originating in Australia and Brazil. More than 50 percent of the ore was being moved to destinations in East Asia and 36 percent to Western Europe. Only 4.5 percent went to North America. With an average length of voyage of 5,494 nautical miles, these are long linkages. For metallurgical coal, the main destinations in 1990 were again East Asia, with 55 percent of the total, and Western Europe, with 28 percent. The single largest importer, at 43 percent, was Japan. The main points of origin were Australia at 36 percent and Atlantic North America at 31 percent. Average length of voyage was 5,436 nautical miles. Also, 41 percent of both cargoes moved in "very large bulk carriers," which is the designation for vessels with greater than 160,000 deadweight tonnes.[60] The same pattern holds today. Brazil's iron-ore exports account for 31 percent of the world total, Australia's for 32 percent. The biggest importers are Asia, with 66 percent of the total, and Europe, with 25 percent. At 41 percent, the largest single-country importer is China. Illustrating the advantages of local supplies, in 2007, imports to Canada and the United States accounted for only 2.4 percent of total world imports.[61]

The shipping industry's unconcentrated structure makes rates highly volatile. International events and changes in the world economy and in demand for ship capacity have a sharp impact on rates.[62] Between August 2002 and February 2004, bulk commodity rates rose by 400 percent, greatly increasing the costs of overseas supply.[63] Together with the 71 percent increase in iron-ore prices the following year – also a result of China's rapid growth – higher shipping rates amounted, in the words of Corus Steel's director of commercial coordination, to a change in the steel industry's supply side. Between May 2005 and May 2006, the ship charter rate from the ore-shipping port of Tubarao, Brazil, to Baoshan, China, rose another 27 percent, from $33,850 per day to $43,000 per day.[64] As was noted in Chapter 2, the ocean shipping differential between Australian and Brazilian ores allowed Rio Tinto in 2008 to negotiate a hefty price premium with Baosteel. China's expansion and continuing reliance on external materials will keep upward pressure on ore prices and shipping capacity, raising costs for all producers that depend on them. The result is a "structural shift in the cost of making steel."[65] Although ore prices fell sharply as the 2008 financial crisis unfolded, the same conditions of supply and demand can be expected to raise them again when the global economy recovers.

To these factors we must now add fuel prices, which began to soar in 2007. In June 2008, world crude oil prices exceeded, in constant dollars, their previous all-time high, which had been set during the oil crisis of 1979-80. That change has directly affected transportation. As was just seen, inexpensive transportation and economies of scale affect industrial location and support trade. The implication of rising fuel costs is directly practical: "In economic geography, the most useful measure of distance is not time or miles but dollars."[66] At 2008-level oil prices, "every ten percent increase in trip distance translates into a 4.5 percent increase in transport costs."[67] The effects are significant: as a result of oil-price increases, transportation costs now exceed tariffs as a trade barrier; at mid-2008 prices, converted into tariff equivalents, transportation costs amount to an average tariff rate of 9 percent.[68] If oil prices recover from their collapse in autumn 2008, shipping heavy commodities such as steel over long distances will become too expensive. That has major implications for the world steel industry, as Chapter 6 will explore.

Despite the distances just seen, the steel industry has not migrated to coal and iron-ore sources in Brazil and Australia. Offsetting factors have favoured existing locations. Sunk costs, established and close linkages with customers, external economies of scale, and location in the largest markets have induced adjustment instead of relocation. The industry's expansion beyond Europe, North America, Japan, and South Korea has been driven by local market growth, and that growth is largest and fastest in China. The prospect is for linkages to develop with steel customers there. As will be seen in Chapter 5, that is presently occurring in the automobile industry. Expansion beyond local demand will produce exportable surpluses. Whether shipping them is feasible (as Chapter 6 will consider) will depend on transportation costs. Here the focus is on how the industry has expanded to its present-day international configuration.

Locale and Trade

Steel is an unlikely export commodity. It has a high unit weight, and except for steel traded within the EU and NAFTA, most of it is shipped across oceans. The same is true of steel's raw materials, which travel the distances just seen. In contrast to the multiple sources of shipping capacity, those raw materials are provided by three large suppliers possessing oligopoly pricing power. And as will be seen shortly, current steelmaking technology provides only very small labour-cost advantages to producers in industrializing countries.[69]

Nonetheless, the present volume of steel trade – more than one-third of world output – shows that large export markets do exist along with willing suppliers.

As suppliers disperse and export, they "invade their competitors' territories, resulting in a new international structure of market shares."[70] The actual results may be moderate if the new imports do not displace the incumbent domestic producers. When those domestic producers continue, the result is more redistribution than displacement, with established centres augmented by newer ones. As these become established themselves, they may generate a two-way trade by importing goods and services, particularly as incomes rise. As a result, "globalization ... may involve a less radical reshuffling of specializations than we have been led to believe by many theories."[71] If high transportation costs do continue, domestic producers will enjoy an enhanced locational advantage – a point that Chapter 6 will consider.

Although low transportation costs have encouraged trade, the result is less dramatic than one might expect. On the one hand, the elasticity of distance as it affects trade volume fell by some 25 percent between 1967 and 1995, encouraging internationalized production.[72] At the same time, if transport costs really were minor, then distance would not figure importantly in the patterns of industrial location. In fact, location still affects trade. An econometric survey of industrial concentration found that "in 1995 a one percent larger distance between trading partners [was] associated with a 0.83 percent lower trade volume."[73] One reason is that factors of production may be much more mobile within countries than between them. Another reason is that, transportation costs notwithstanding, the forces maintaining particular industrial regions may be significant.[74] In light of what has just been seen, likely factors are sunk costs in large installations and in the linkage networks that support external economies of scale. Also germane is the influence of informal commercial networks that favour local producers. Chapter 6 considers the effect of network factors on the world steel industry's future organization.

A different pattern may arise when transportation costs for intermediate goods are cheaper than for final goods. When intermediate goods are less expensive to ship, producers can supply from more remote locations.[75] That has been true with steel. Efficient producers with immediate access to ocean shipping terminals have developed a strong export trade, as will be seen presently with Japan, South Korea, and Brazil. Russia, in contrast, depends on rail transportation to its European markets, and long overland distances separate it from its ocean shipping point on the Pacific. As will be seen presently, however, very low production costs in Russia offset that disadvantage. More strikingly, a trade in semi-finished steel has developed, with Brazilian, Russian, and Ukrainian producers supplying mills in North America, Europe, Japan, and China. Patterns such as these may make industrial location and trade more diverse and interdependent than one might expect: "Trade and trade theory ... are not mirrors of location. They are, variously, complements, outcomes, and partial causes of location." Research on globalization must "[measure] international location in its own right rather than [read] it off from trade patterns."[76]

A useful typology illuminates location and trade by classifying industries along two axes. The first axis is territorialization – the degree to which production systems must be situated in particular places. The high end of this axis indicates production that is centred in particular regions, and the low end indicates production that is easily dispersed and that requires no territorial core. The second axis is the internationalization of production – the degree to which industrial output is sold on world markets. The high end of this axis indicates production that is sold internationally, and the low end indicates production that serves local markets. The integrated steel industry fits the High/High quadrant, with fixed and specific locations of production and a product that can be exported to world markets.[77]

Some aspects of this typology pertain to the steel industry quite strikingly. Firms can move into the High/High category as they "gain the ability to market their products around the world without losing their locationally specific assets." In the steel industry, as was seen in Chapter 2, the required concentration of expensive and high-output capital makes these assets locationally specific indeed. Required for global marketing is internationalized demand.[78] Steel's demand derives from its wide use as an intermediate good. Two related conditions are liberalized trade, which opens international markets, and transferable technology, which allows production to be set up at new sites and for new locales to be formed. In this way, new centres of international production can emerge "via what might be called the ongoing reinvention of relational assets in the context of high levels of geographical openness in trade and communication." Altogether, several elements stand out: the importance of localized assets and external economies of scale, product specialization, and competition among large firms – referred to by Storper as global supply oligopolies.[79]

Transferable technology is particularly worth noting. The capital equipment for steel mills is highly standardized, made by a few firms, and for sale globally. The advantage is that enterprises or governments seeking to enter steel production or to upgrade old facilities do not have to take on expensive design and manufacture. Instead they can buy and assemble available components. When they do so they import current knowledge along with current technology. That confers the ability to be cost-competitive with established producers – more so if their own plants are less up-to-date and if transportation costs allow their home markets to be served.[80] Another way of transferring technology is through joint ventures in which the partner provides the means of making more advanced products. That pattern is visible in the European and Japanese joint steel ventures in China, where technology and knowledge for high-end products is being adapted to facilities that are themselves new. Chapter 6 examines technology transfer as one way for makers of commodity steel to advance into higher-end products.

One element of the typology that did not fit the steel industry until recently is international ownership – a pattern that could be subsumed by the term "global supply oligopolies." Steel producers had been nationally based and often government owned, but as was seen in Chapter 2, in Europe and the United States the industry restructured and consolidated – under government direction and sponsorship in Europe with the melding of three formerly state-owned producers into Arcelor, and under corporate acquisition and restructuring in United States as International Steel Group and United States Steel Corporation salvaged viable parts of bankrupt producers. The process has become global: Mittal Steel, which began by acquiring mills in Indonesia, Trinidad, and Mexico, has expanded its holdings into North America and Europe. When Mittal, having acquired Inland Steel in 1998, merged with International Steel Group in 2005, it became North America's largest producer; and when it acquired Arcelor in 2006, it became the largest producer in Brazil, where Arcelor had become a major presence. Mittal and Arcelor were the world's two largest steel producers, and their merger triggered expectations that the industry would consolidate globally. There are considerable incentives for doing so, as Chapter 5 will examine.

Another element of the typology that does not fit is product differentiation and export potential. Offering similar but not identical goods would indeed allow formerly localized producers to compete against established incumbents abroad. But steel – especially long products, hot-rolled coil, and plate – is a globally standardized and highly substitutable commodity, and there is nothing intrinsic about it that makes buyers exclude an alternative supply. When that indifference is present, as was seen in the previous chapter, price and availability enable entry into markets previously dominated by national firms. The exception is at the top end of the line, where product variety is "notoriously complex" and where automakers may draw on a sophisticated and finely specified array.[81] Underlying this differentiation are the rapid evolution of car body design and increasingly high standards of strength, corrosion resistance, and lightness. Steelmakers can either work closely with particular automakers, or they can develop products and technology for the whole industry. Chapter 5 examines their efforts.

New Producers
For developing countries, chronic shortages of foreign exchange were an incentive to construct basic industries such as steel. Producing domestically would conserve the foreign exchange needed to purchase other imported commodities. But the technology itself had to be imported, and that expense

had to be weighed against alternative investments and their utility for national development. For capital-intensive industries especially, the investment costs could easily outstrip domestic financing. These considerations persuaded the World Bank to support steel-industry investment in developing countries, and in the 1970s it was a primary source of funds for Brazil, Turkey, and Mexico. The bank's support encouraged manufacturers of capital equipment to contribute financing of their own. That combination "was decisive in getting many steel projects off the ground in the periphery."[82]

The original purpose of building steel industries in South America, South Africa, and India was import substitution, a development strategy popular at the time whereby states sought to free themselves from the pattern of exporting raw materials and importing finished goods. Besides saving foreign exchange by producing manufactured goods domestically, the strategy was expected to generate forward and backward linkages that would encourage more general industrial growth. Those linkages were necessary to avoid one of import substitution's major pitfalls – industrial development limited to high-cost and inefficient consumer-goods industries producing for small home markets.[83] The Soviet Union and China pursued similar objectives under their industrialization, and their goal of economic self-sufficiency had elements in common with import substitution. In both countries, steel industries were intended to supply the other sectors.

Forward and backward linkages are indeed abundant in modern industrial economies. In Canada, an input-output table study of the impact of the primary-metals sector found that the industry's backward linkages – which reflect the impact of supply and investment – generated 1.77 percent of GDP, and that the motor-vehicle industry was the only sector with a larger contribution, at 2.21 percent of GDP. The same was true of forward linkages. Canadian mining and primary metals (aggregated in that section of the survey) had the highest number of industries for whom purchases from the sector accounted for more than 10 percent of their total input purchases.[84] Achieving that fabric of production is one of industrial development's most appealing prospects.

Factors favouring steel industries in new locales include these: (a) available current technology, (b) cheap and readily available energy or raw materials, (c) favourable levels of pollution regulation, and (d) domestic economic growth that is generating demand for capital and infrastructure and for consumer-goods industries.[85] Introducing new technologies into countries that are less developed than the world's mature industrial economies requires a role for the state, either directly as a promoter and planner or indirectly as a provider of the conditions for adopting technical knowledge and best practices – a process that involves a variable and complex set of relationships.[86] An important benefit of rapid development in an industry is that it encourages mastery of the related technical knowledge and procedures as well as a quick move along the learning curve.[87] Cycles of expansion and retrenchment in particular industries provide opportunities for such transfers, particularly in industries, such as steel, that have very high capital costs. During low points in the cycle, incumbent industries, facing declining profits and wary capital markets, may defer or forgo new technologies, providing newly establishing industries elsewhere with an opportunity to secure cost and productivity advantages.[88] Factors (a) and (b) favour producing steel for export, and adopting the latest technology – especially when incumbent producers are not doing so – favours export advantages.

As was seen earlier, the standard theory of comparative advantage may not account for the distribution of particular industries. Instead, specialization and economies of scale may generate efficiencies and trade advantages. The state can play a direct role in that process – in this regard, the South Korean and Brazilian steel industries began as state enterprises. The same was true of Russia under the Soviet Union and much more recently of China. Through either subsidies or direct ownership, governments are able to foster industries in locations that market forces would not otherwise have selected.[89] By "distorting relative prices," government involvement offsets incentives that would make different investment allocations and shapes the global distribution of industries.[90] When these investments are large and when they are concentrated in industrializing countries, they rearrange the spatial organization of world industry.[91] When a sponsored industry develops export advantages through specialization and economies of scale, much of the credit must be allocated to state intervention. When transportation costs to overseas markets are favourable and overseas trade develops, the process fits the pattern seen earlier, with economies of scale allowing locations away from existing market clusters.

Becoming efficient by expanding production suits the steel industry's economics and technology, which require very large operations. High volume and specialization give steel producers in industrializing countries the same advantages of scale as those in the industrial economies.[92] For some developing economies, the necessary scale may be too large. When that is the case, output levels exceed absorptive capacity, and the necessary degree of specialization limits output to a narrow range. Producing a fuller product line at that scale simply creates a more diverse domestic oversupply.[93] Government motives may intervene. When governments pursue regional development or labour market objectives, or more generally when they sponsor new industrial clusters, their involvement may extend to supporting set levels of output in order to prevent layoffs and maintain the investment's rationale. When they do so in the face of falling domestic or international demand, their action amounts to subsidizing surplus capacity and production.[94]

Exporting was not the original motivation for expanding and modernizing Japan's steel industry and for building new steel industries in South Korea and Brazil. All were initially concerned to provide a domestic and inexpensive source of steel for their growing manufacturing sectors. Seeking economies of scale, Japan and South Korea soon developed surplus capacity for export. Additional encouragement came from their governments. In the 1950s, Japan's Ministry of International Trade and Investment helped arrange financing for the early stages of steel expansion, and in the 1980s the governments of Brazil and South Korea were "especially aggressive" in expanding steel production. The need for their involvement was directly related to steel technology, which requires a very large and concentrated initial investment. Achieving economies of scale and efficient operating rates requires a large market to absorb production. When that output exceeds domestic markets, the surplus must be exported.[95]

World market demand affects product specialization. When oligopolistic production characterizes an industry internationally, as it does steel, production will tend to be "in accordance with the tastes of the larger market," whose demands form the "product space" to be filled.[96] Any steelmaker can fill that product space with an internationally standardized commodity. High substitutability removes the barrier of product differentiation, and export sales, as was seen in Chapter 3, can be generated through price discrimination and discounting. That practice, as was also seen, can generate trade complaints. Less purposefully, episodes of global oversupply put downward pressure on prices, and the least efficient producers are the first to suffer. The effect is magnified by the highly cyclical nature of steel demand and by the product's price volatility.

Scale Economies and World Markets

The Japanese, South Korean, and Brazilian steel industries all pursued economies of scale and became high-volume producers. Although the three began with the objective of providing steel for growing domestic industries, they soon were producing surpluses and began exporting. A brief look at these three industries shows how the basic economics of steel production led to world markets and to varying patterns of success. A look at the Russian steel industry shows how that centrepiece of the Soviet system privatized and became the world's fourth-largest producer and fourth-largest exporter. The section concludes with a review of China's steel industry, whose rapid expansion heralds an impressive – and potentially disruptive – entrance into the world market.

Japan's steel industry followed a model of export-led growth that is by now widely familiar, as is South Korea's. The Chinese government's movement away from state ownership can be explained readily in reference to steel. These considerations merit a fairly concise treatment. On the other hand, the Brazilian and Russian steel industries have enough novel elements to require fuller attention. They are also, in the author's view, the most interesting.

Japan

Japan established a successful model of steel production in the 1950s when it began modernizing and expanding its industry by adopting the then-new technologies of BOF and continuous casting. Although the original purpose was to supply growing postwar industrial production and eliminate steel imports, modernization and expansion produced a strong export advantage. To secure the increasing returns of scale that were becoming possible with modern steelmaking technology, the major Japanese steelmakers, with financial backing from the Ministry of International Trade and Investment, pioneered the development of high-capacity blast furnaces and continuous-production processes. The most important form of aid was an export subsidy, which was begun in 1953 and which was terminated when Japan joined GATT.[97]

Because they were dependent on overseas supplies of both iron ore and coal, Japanese steelmakers negotiated long-term contracts and, with the government's backing, helped finance Australian iron-ore production. To reduce the costs of long-distance sea transportation, the steelmakers, the shipbuilders, and the government developed a trilateral arrangement that led to the design and production of a new generation of high-capacity bulk carriers.[98] A special shipping company was established to handle transportation, and new mills were built at seaports. Cheap raw materials and efficient current technology were Japan's advantages on the world market. Exporting was made necessary by two requirements: to pay for ore and coal imports, and to absorb the output of the new high-capacity mills. By the late 1970s, those mills were the non-communist world's largest.

By that time, Japan's manufacturing industries were mature and their demand for steel had levelled off. Because of their scale and volume, the mills had no choice but to export some one-third of production.[99] As that change was occurring, world demand stagnated, and the Japanese steelmakers, along with their counterparts in Europe and North America, faced surplus capacity.[100] At the same time, Japanese wage rates began to erode cost advantages. The steelmakers sought to expand market share by price discounting and meanwhile were forced to retrench. That involved adjustments similar to those in the United States: reductions in capacity and workforce and specialization in higher-end products.[101] Added to these conditions were the move of Tokyo Steel, Japan's largest minimill, into higher-end steel products; the emergence in South Korea of a powerful steel industry, which began undercutting prices in Japan's domestic market; and the recovery of steel producers in the EU and North America. Although Japanese steelmakers had reduced their labour forces by 40 percent since the 1970s, by the mid-1990s they faced further restructuring.[102] A major event was the merger in 2002 between Kawasaki Steel and NKK Corporation to form JFE Holdings, currently the world's third-largest producer behind the merged ArcelorMittal Steel and Nippon Steel. Demand has come from exports and from a recovering Japanese economy, in which production of automobiles and electronic goods has risen to record levels. Japanese manufacturers have also increased their demand for steel as they have upgraded their production facilities with new machinery. Significant as well are increased steel exports to the world market and decreased steel imports from China. Japanese steelmakers have used these returns to develop new products and to improve their facilities. Steel production reflects these developments: 2007's output of 120 mmt exceeded the previous record of 119 mmt that had been reached in 1973 during the industry's peak of postwar expansion.[103]

South Korea

The government of South Korea sought to emulate the Japanese steelmakers' success. Because of insufficient domestic investment capital, the government played a central role in assembling the necessary financing to establish the Pohang Iron and Steel Company (POSCO) in 1968.[104] Originally owned by the government, the company has been privatized, with the government currently holding 3 percent of the shares. POSCO's initial role was to produce steel for Korea's expanding manufacturing industry at a price below imports. Korea's manufacturing development was based on export growth, and POSCO's role as a supplier of intermediate goods was part of that broader scheme. Between 1962 and 1999, domestic consumption of steel rose from 0.3 million to 39.9 million tonnes, while steel production rose from 0.1 million to 41 million tonnes.[105] In 2007, steel consumption was 54.8 mmt and steel production was 51.5 mmt. In 2006, the most recent year for IISI import and export figures, South Korea, with steel exports of 18 mmt and imports of 22.4 mmt, was a net importer.[106]

Like Japan's steel companies, POSCO relied on cutting-edge technology, high efficiency, and volume to compete in the world market. And like Japan, South Korea set an export target at 30 percent of production. Japan had been South Korea's main external steel supplier, but by 1981 the amount of Korean steel exported to Japan had begun to exceed the amount imported from that country.[107] Like Japan's steel industry, South Korea's was based on imported raw materials, and POSCO's second mill, begun in 1981, was built at a seaport. As did Japan, POSCO reduced the cost of imported raw materials through long-term contracts and by investing in overseas mineral development, although POSCO still depends on purchased ore imports and must negotiate prices yearly. In 2008, POSCO, along with Japanese steelmakers, had to accept a 65 percent price increase from Vale, although soaring steel demand in Asia allowed it to raise its own prices.[108] South Korea and Japan were industry leaders in automating production and computerizing process management.[109] In the early 1990s, South Korean steel production expanded further by adopting minimill technology; output from that sector climbed from 8.4 mmt in 1990 to 17.8 mmt in 1995 and to 24 mmt in 2007.[110]

Among the world's steel producers, POSCO currently ranks as fourth-largest, with a 2007 output of 31.1 mmt.[111] Its advantages include highly modern and efficient plants, large economies of scale, a capacity utilization rate of 100 percent, and a non-union workforce, giving it one of the world steel industry's lowest cost structures.[112] Compared to North American steelmakers, POSCO is more profitable than Nucor, the most profitable producer. To improve its access to the American market, POSCO has entered into a joint venture with United States Steel, a move that has enabled it to evade many of the protective effects of the special steel tariffs imposed by the Bush administration in 2002. More about that joint venture will be seen later; the Bush tariffs will be examined in Chapter 5. South Korea's manufacturing sector has also grown, and domestic sales account for two-thirds of POSCO's production.[113] POSCO intends to expand domestic capacity to 40 mmt over the next ten years. Beyond that, further expansion will be in foreign operations, with a goal of 20 mmt by 2017.[114]

Meanwhile, POSCO's cost advantages are being challenged by modernization and productivity increases in North America and Europe and by even newer mills in China, whose development has drawn from POSCO's own technology and from its management and engineering experience. South Korea was China's largest export customer in 2005; the 6.7 mmt volume recorded that year represented a 52 percent increase over the previous year.[115] To address the surging demand in China, POSCO is seeking a joint venture there – a move that is "strategic rather than profit-making." Its ideal partner would have an annual production of around 20 mmt. (For comparison, United States Steel's 2007 output was 21.5 mmt.) POSCO also intends to expand into India. In the words of POSCO's chief financial officer, "the domestic market has matured so we have to find growth abroad. We will seek organic growth through greenfield investments in immature markets while looking for M and A chances in mature markets."[116] If growth in immature markets falls below expectations, POSCO plans to use those facilities to supply slabs for the home or export market.[117]

At home, POSCO's market is beginning to resemble those of its European, Japanese, and North American counterparts. The Korean economy is reaching a plateau of manufacturing output, and its demand for steel will be growing more slowly. As well, minimill expansion threatens to oversupply POSCO's domestic market. Finally, compared to industry leaders such as ArcelorMittal and Nippon, POSCO's production efficiency has not been matched by the same level of product development – a shortcoming that affects the crucial higher end of the line. In 2007, POSCO and ArcelorMittal held exploratory talks involving POSCO's paid use of "special recipes" for automotive steels. In return, ArcelorMittal hoped to benefit from POSCO's presence in China.[118] Unlike European and North American steelmakers, POSCO is outside the EU and NAFTA; for that reason, it faces export disadvantages and deliberate trade restrictions.[119] Even so, soaring steel demand in Asia, together with the Chinese government's export restrictions to avoid domestic steel shortages, enabled POSCO to earn a record profit in the second quarter of 2008.[120]

Over the longer term, the most promising path for South Korea's steel industry is the same one as for North America and the EU: shifting to higher-end products and achieving further efficiency. This is true not only for POSCO but also for the smaller Korean steelmakers, Inchon Iron and Steel and Hanbo Steel. Hanbo went bankrupt in 1997 amidst the developments that produced the Asian economic crisis and was kept operating with government aid. Both firms were purchased by the Hyundai group to form a new steel production unit, Hyundai Hysco; this has enabled Hyundai Motors to become self-sufficient in steel. In 2007, Hyundai ranked twenty-ninth among the world's top eighty steel producers. This development has not pleased POSCO, which – its own expansion notwithstanding – fears excess capacity.[121]

Brazil

Steel production in Brazil began under state ownership as a designated strategic industry. An important goal of this was to make industrial use of one the world's two largest deposits of high-grade iron ore. (Another important goal was import substitution.) The first integrated mill was built in 1948, and by the 1990s the number had grown to thirty-five. The government owned 70 percent of the steel industry until privatization began in 1988. An exception was the minimill Gerdau SA, which had always been privately held. To provide reasonably priced steel to domestic industries, the government regulated steel prices and absorbed the financial losses. As manufacturers, particularly the automobile industry, began exporting in the 1980s, the government's rationale for regulating steel prices expanded to include international competitiveness.[122]

Exporting did not begin as an unintended by-product of expanded scale. Rather, it was necessary in order to offset low earnings caused by government domestic price controls, which were set well below international levels. Since those domestic prices were lower, Brazil's regulation did not fit the model of international price discrimination seen in Chapter 3. In that model, governments protect domestic markets to allow producers to price-discriminate by charging high prices at home and offering discounts abroad. Instead, Brazil's rates of protection for steel were actually below those for the domestic manufacturing sector.[123] After the world oil crisis of 1974, an additional export motive was to earn foreign exchange to pay for oil imports.[124]

The integrated sector was afflicted by high costs, labour unrest, and delays in adopting new technology, particularly continuous casting.[125] Privatization, along with trade liberalization, was part of a program of national economic reform, and steel, along with petrochemicals and fertilizer, was in the first wave.[126] The move has been seen not as bending before the forces of neoliberal reform but as a coherent strategy pursued by a relatively autonomous state.[127] In 1997, the government also privatized Companhia Vale do Rio Doce, which owns the majority of Brazil's iron reserves and is the world's largest iron-ore producer. Vale, as the company is now known, also owns a railway, which moves the ore cheaply to Brazil's mills, and an ocean terminal facility.

Consolidation through mergers has reduced the industry from thirty-five producers to nine, of which four have predominated: Usiminas (formerly Usinas Siderurgicas de Minas Gerais), CSN (formerly Companhia Siderurgica Nacional), CST (formerly Companhia Siderurgica de Tubarao), and Gerdau SA.[128] As in the United States and Europe, reform and restructuring have been directed at raising efficiency and profitability by integrating the most productive components into a few large firms. Closing old mills and production lines removed some 2.2 mmt of capacity; meanwhile, $12 billion of investment was added in metal-making and rolling plant. The major producers reduced their workforces by 26.6 percent. The most dramatic productivity improvement (48 percent) was at CST. Usiminas' and CSN's improvements were smaller because their previous operating practices had kept them much closer to their optimal efficient scales.[129]

Even so, those improvements have been seen as a "correction of past inefficiencies rather than an extension of the global envelope of competitiveness."[130] And unlike Japan and South Korea, Brazil did not develop technological and process innovations. Instead, these were imported, along with new mill and control equipment. There is now some latitude for in-house research and development; that said, external dependence, according to a recent assessment, raises "serious questions ... surrounding the current and future course of national self-reliance in steelmaking technologies."[131] Also unlike Japan and South Korea, Brazil has financed its steel industry through heavy borrowing abroad.[132] The high interest rates on commercial debt previously limited the gains from modernization, although the Brazilian currency's much more favourable present exchange rate has reduced these effects.

Brazil's rich ore supply remains a basic advantage. Together with Vale's efficient extraction and rail delivery, this advantage provides ore to Brazilian mills at prices well below delivered prices abroad. One index of that supply's sheer size is that in 2007, notwithstanding its status as the world's ninth-largest steel producer, Brazil exported fully 77 percent of its produced ore.[133] The same advantage is true for electricity costs for Brazil's minimills. Steel scrap is scarcer and more expensive in Brazil than in North America. However, only a tiny imported amount – 0.1 mmt – was required in 2004 to supply a consumption of 12 mmt. In 2007, Brazil was able to meet all of its scrap requirements domestically.[134] Overall, this amounts to a factor abundance that favours steel production – an advantage that is reflected internationally in Brazil's historically very low level of steel imports.[135] In 2006, Brazil was the world's fifth-largest net steel exporter, at 10.7 mmt.[136] These advantages do not include labour costs (see below), but they do buttress the notion that production efficiencies support remote supply sources for intermediate goods in general. For Brazil, those goods are weighted to semi-processed steel. In 2003, Brazil's costs of producing steel slab were around half those faced by the United States and Europe.[137]

Even in areas in which there appear to be strong advantages, however, cost differentials can change. In 2007, a rising Brazilian currency, a low American dollar, and burgeoning demand in Brazil from the automobile and petroleum industries led not only to a steel shortage but also to attractive prices for imported steel. Importers include the Brazilian steelmaker Usiminas, which dominates the flat-steel sector, and the Ford Motor Company. For Usiminas the product is heavy plate; the suppliers are China, the CIS, and the United States; and the price for delivered steel – even at high ocean-shipping rates – is only 8 percent above the f.o.b. price of Brazilian steel. For Ford the product is automotive steel; the source is the United States; and the motive is lower costs and scarcity. As in China, rapid economic growth is putting pressure on domestic steel supply, with the result that "if Brazil is to enjoy economic growth of more than 4 percent a year and still export some steel, large-scale steel imports are here to stay."[138]

CSN and Usiminas operate regular integrated facilities and produce a standard array of hot- and cold-rolled long and flat products. In 2007, Usiminas, with an output of 8.7 mmt, ranked thirty-fifth among the world's steel producers, and CSN with an output of 5.3 mmt ranked fifty-ninth.[139] In contrast, CST – ranked sixty-sixth at 3.8 mmt before its 2005 merger with Arcelor – concentrates on slab production and exports one-third of it to North America. It has also invested $2.2 billion to expand capacity with two additional blast furnaces and to move into higher-end steel production with a new rolling mill.[140] For its part, Usiminas in 2008 announced a four-year, $14 billion plan for a new slab mill; it intends to export 60 percent of its output.[141] Together, those three producers turn out just over half of Brazil's steel.[142] Much of Brazil's steel production is in long products – beams, rods, and wire – and, as in North America, minimills are major producers.

In 2007, Brazil was the United States' fourth-largest supplier, behind Canada, China, and Mexico.[143] It is the third-largest source, behind Mexico and Canada, of semi-finished steel.[144] The American mills' strategy is to concentrate on higher-margin products at the upper end of the product line and leave lower-end items such as structural steel to minimills and foreign producers.[145] This is fully in keeping with qualitative forms of market adjustment in which producers shift to more advantageous products.[146] As these mills abandoned primary steelmaking they created a demand for slab and hot-rolled coil.[147] Between 25 and 30 percent of the steel imported into the United States is used by domestic mills.[148] Ohio-based AK Steel's newest mill, built to produce high-end products for the automobile industry, uses steel from the company's own blast furnaces but is readily convertible to foreign slabs.[149] Canadian producers use imported slab as well, although their supplier by a wide margin is the United States. In announcing greatly reduced third-quarter earnings for 2005, Dofasco cited as one reason a large inventory of imported slab that had been purchased in late 2004, when world steel prices were rising rapidly. The price difference between slab and finished steel added more than $40 million to Dofasco's costs, which rose from 80 to 91 percent of sales.[150] These offshore supply relationships all work through regular export channels. More direct arrangements have evolved in two forms.

In the first, foreign steel firms merge with Brazilian producers or enter joint ventures. In 2002, Corus, formed from the 1999 merger of British Steel and the Dutch steelmaker Hoogovens, announced a planned merger with CSN; the goal in this was to secure an overseas slab supply, concentrate on high-end products, and escape dependence on the world iron-ore oligopoly.[151] CSN's attractions were its own iron-ore mine and a modern and efficient mill. Corus' abandonment of the plan four months later came not from a strategic change but from its own operating losses and uncertainty about Brazil's economy.[152] In 2004, South Korea's Dongkuk Steel, the Italian metals group Danieli, and Vale announced a project for a new Brazilian mill to produce slab for export. In the same year, Baosteel, China's largest producer, announced a prospective joint venture with Vale for a new blast furnace and mill to produce slab for China.[153] Costs were the motivator: with slab being much cheaper to ship than raw iron ore, building a slab mill in Brazil was viewed as a more efficient option.[154] Baosteel abandoned the project two years later because of construction costs and an inability to reach agreement with the government of Brazil's Maranhao state about location. In 2007, however, it announced a second venture with Vale in the state of Espirito Santo for a 5 mmt slab mill to be opened in 2011.[155]

Joining iron-ore supply with high-end production has also involved Russian and European steelmakers. To secure a slab supply from a producer with access to local iron ore, Corus in 2006 entered exploratory merger talks with the Russian steelmaker Evraz. In 2004, two of Russia's wealthiest steel magnates acquired a stake in Corus. Speculation at the time was that they hoped to integrate one of the firm's facilities with their operations by supplying slab from their Russian mills and transferring Corus' technology.[156] Supply calculations also figured in United States Steel's 2007 acquisition of Stelco. Besides sheet steel for the automobile industry, Stelco produces slab, and gaining that nearby source made the company doubly attractive.

A more comprehensive strategy has been pursued by Arcelor. To secure a source of slab, Arcelor acquired part ownership of CST in 1998. By 2004, CST had become the world's largest slab exporter.[157] With an eye to Brazil's domestic market, Arcelor also acquired Acesita, a maker of stainless products. In 2004, Arcelor expanded its stake in CST to majority shareholder and, in a joint venture with CST and Vale, opened a new steel facility, Vega do Sul, to produce sheet steel for Brazil's automobile industry. In 2005, Arcelor announced that CST, long-product maker Belgo, and Vega do Sul would be merged into a single entity, Arcelor Brasil, whose combined capacities would make it Brazil's largest steel producer.[158] Brazil's economy provides an expanding domestic market, with booming construction and automobile industries creating a demand for finished and specialty steel that amounts to 60 percent of domestic production. Arcelor plans to double production at one of its main Brazilian plants by 2012. Since being acquired by Mittal Steel, Arcelor's plans have become part of a larger composite, and an important asset in the purchase was Arcelor's carefully developed position in Brazil. There is indeed potential in the Brazilian market. Now that the government no longer regulates prices, Brazilian steelmakers prefer the domestic market for its higher profit margins.[159]

In the second kind of relationship, Brazilian producers acquire a stake in American mills. One of the largest purchases was of defunct Kaiser Steel by Vale and Japan's JFE. Each has a 50 percent stake in the new company, California Steel Industries. The acquisition involved scrapping the steelmaking facility, located east of Los Angeles, and using the rolling mill to process Brazilian slabs into high-end products.[160] In a similar move, CSN purchased bankrupt Heartland Steel in Indiana in 2001, enabling CSN to supply slabs for Heartland to finish into specialty steels. According to a press announcement, "this is an important step in the internationalization of CSN."[161] In the European market, CSN acquired full ownership of the Portuguese steel producer Lusosider in 2006 with the stake previously held by Corus. Buoyed by good profits and available cash, Brazilian producers are expected to make more foreign acquisitions.[162] A step down in the production process, Vale is shifting production of iron pellets (the processed feedstock for blast furnaces) from Brazil to Mozambique to take advantage of cheap local coal supplies (Brazil imports its metallurgical coal from the United States, Australia, and Canada), using iron ore shipped from Brazil.[163]

An important joint venture in the United States uses slab from South Korea. United States Steel had previously supplied its plant near San Francisco with unfinished steel from a mill in Utah. Aging plant in both facilities and high transportation costs made the operation uncompetitive with imports. US Steel closed the Utah plant and, using financing from POSCO, upgraded the plant near San Francisco to finish POSCO-supplied steel. These arrangements can also be constructed under a single corporate aegis. Mittal Steel owns both Inland Steel in the United States and Ispat Mexicana SA de CV. Slab production costs in Mexico, according to a Mittal executive, are among the world's lowest, making it an ideal supplier for Inland's finished products.[164] One cost advantage is Mexico's abundant supply of gas and oil, which favours DRI production.[165]

On the electric-furnace side of the industry, Brazilian minimills compete less with integrated producers in North America than with domestic minimills, whose market emphasis is in the same products. Brazil's Gerdau SA, through its subsidiary Gerdau Ameristeel, owns sixteen minimills in the United States and three in Canada, making it North America's second-largest minimill operator after Nucor and the largest producer of long products in the hemisphere. Pursuing a strategy of integration, Gerdau merged its North American operations with the Canadian minimill Co-Steel Inc. in 2002 to become Gerdau Ameristeel Corporation, with headquarters in Toronto and Tampa. In 2007, Gerdau Ameristeel purchased Chapparal Steel, North America's second-largest structural-steel producer. Parent firm Gerdau's production abroad now exceeds home production in Brazil. On the integrated side of the industry, Gerdau had been evaluating Dofasco as an acquisition before it was purchased by Arcelor.[166]

Compared to Japan and South Korea, Brazil's abundant supply of high-grade iron ore gives it a locational advantage, enabling it to establish a "significant position in the world market for semi-finished steel products."[167] Unlike Japan and South Korea, however, Brazil relies on purchased foreign technology. Notwithstanding the top-level technology and metallurgy now available through ArcelorMittal, that dependence raises the question of whether Brazil can ascend the product line into higher-end and specialty steels. Will it remain predominantly a supplier of "metallic inputs" for producers abroad?[168] Chapter 6 considers motives for moving up the industry's hierarchy.

Russia

Large-scale production of pig iron began in the Urals under Peter the Great and was based on high-grade and readily accessible ore and abundant charcoal. For a time during the mid-seventeenth century the Urals were the world's largest iron-producing area and exported to Great Britain. With the advent of steelmaking technology in the mid-nineteenth century, production centred in the Donets Basin in eastern Ukraine, where massive coal and iron-ore deposits were located in ideal proximity. Under the Soviet Union's heavy industrialization, steel production was developed in the Urals to take advantage of the vast Kuznetsk coal deposits in western Siberia. Massive mills were built in the Kuznetsk basin itself, and the three regions – Urals, Kuznetsk, and Donets – became the country's largest steel-producing districts. Subsequently, the Soviet government expanded steel production in Chernopovets, northeast of Moscow, and in Lipetsk in western Russia. Because the Donets is the only region where coal and iron ore are located near each other, steel production in Russia has always involved lengthy logistical connections between raw-material sites, steel mills, and markets.[169] This remains true today: some Siberian steel is too remote from Europe to be cost-competitive with other Russian producers but much better located to serve markets in Asia. With the collapse of the Soviet Union in 1989, Ukraine's large steel industry became a foreign supplier, and at one point in the 1990s the Russians accused Ukraine of dumping.

The Soviet Union produced prodigious quantities of steel, partly because the government emphasized machine and weapons production, and partly because steel use was extraordinarily inefficient. Poor quality and the wrong varieties made much steel unusable, and poor materials control in manufacturing produced abundant scrap. The unusable steel, often after long periods of storage at manufacturing plants, would be returned to the mills, melted, and remade into the same products. Estimates of waste ran to 50 percent of output.[170] The high point of steel production was the year before the Soviet Union's collapse, when it reached the level of 160 mmt, but by 1995 steel demand had plummeted by almost 50 percent. Manufacturing had collapsed, and in the single year 1994 factory output fell by a staggering 45 percent. Declines in the machine industry were particularly drastic, but arms manufacture and new construction were also sharply down. The ending of price controls under economic reform cut steel demand further when steelmakers promptly raised prices and their customers, struggling under collapsed demand for their own goods, cut their purchases.[171]

Privatization of steel producers began in 1992 when state mills were converted into joint-stock companies. Of the emerging new firms, OAO Severstal, which took over the facilities in Chernopovets, began on a strong note by modernizing its plant and restructuring its labour force. Other firms continued to operate outdated mills and keep full labour complements. Energy discounts and tax credits were provided to forestall layoffs and unrest, but these measures sustained excess capacity and obsolete facilities. One-quarter of the mills used open-hearth furnaces whose efficiency was only one-tenth that of American mills. The result was a market that was "saturated with cheap, bad quality steel and inefficient producers."[172] The most successful managers were those who could negotiate subsidies and discounts, and there was little incentive for firms like Severstal to invest.[173] The steelmakers shared the financial predicament of European and North American steelmakers in the 1960s and 1970s: depressed markets, meagre revenues, and an acute need for modernization. The Russian situation was even worse because there was no adequate financial market; this forced producers to modernize with whatever resources they had.

Russian investors, many of them former directors and managers, acquired blocks of shares in the new enterprises. There was some outside participation. Initial investors in Novolipetsk Steel, formed from the state-owned Lipetsk mill, included George Soros and the Harvard University endowment fund, and foreigners held 40 percent of the firm's shares. After a legal battle to secure representation on the board, the large holders sold out in 1999 and left the firm's director with majority control.[174] Less conventional means of acquisition involved using personal connections with other enterprises and the government to cut off supplies to rival firms or to manipulate shares in takeovers.[175] Aleksander Abramov added two enormous Kuznetsk mills to his original holding of a large Urals mill by purchasing their debts, suing for bankruptcy, and then buying the shares.[176] He combined the three facilities into Evraz Group S.A., which joined Severstal, Magnitogorsk Metallurgical Combine, and Novolipetsk to become the core of the Russian steel industry.[177] The industry consolidated, and by 2007 these four firms, plus Metalloinvest – whose primary business is iron ore but which has large steel holdings – accounted for 80 percent of Russia's output.[178] Although consolidation was accomplished in "questionable" ways and concentrated ownership among a group of industrial oligarchs, it did produce "a handful of global competitors headed by tough and savvy owners and managers."[179]

After reaching a low point in 1995, Russian steel output began to recover. The reason was not a revival of domestic demand but a turn to exporting – a venture strongly assisted by a depreciated rouble. Initial exports rose rapidly and nearly doubled between 1992 and 1994.[180] In addition to a favourable exchange rate, Russian exports benefited from low labour, energy, and materials costs along with the bonus of the facilities having been acquired cheaply. Additional assistance came from the European Bank for Reconstruction and Development, which granted Magnitogorsk a $105 million "lending facility" to help it position itself for foreign markets.[181]

With the rouble's collapse in the currency crisis of 1998, Russian steel exports surged, triggering conflict with Europe and the United States. The

EU had already imposed restrictions on Russian steel imports and during the currency crisis refused to accept greater levels to aid Russian recovery. The United States charged that the steel was being diverted to the American market, which absorbed an amount four times greater than the EU's.[182] For its part, the United States was preparing an anti-dumping case that would involve duties of 100 percent. To avoid proceedings, the two countries negotiated an agreement by which Russia would halt hot-rolled steel exports to the United States for six months and in sixteen other steel categories raise prices and cut volumes. The result was a 68 percent overall reduction for 1999.[183] Exporting grew nonetheless, and returns were augmented by the revival of the Russian petroleum industry and a recovery of construction. For the domestic steel market more generally, transportation costs provided a measure of natural protection.

World demand began to surge in 2002, and Russian steelmakers shared fully in the proceeds. By 2007, Russia ranked as the fourth-largest net exporter behind China, Japan, and Ukraine. Severstal achieved the position of the world's fifteenth-largest producer, Evraz the seventeenth, Magnitogorsk the twenty-first, and Novolipetsk the thirtieth.[184] These four had accumulated an estimated $8 billion in free cash, with the largest portions going to Magnitogorsk, Severstal, and Novolipetsk.[185] Some of the returns were used to buy iron and coal mines in Russia, so as to achieve vertical integration and provide a buffer against commodity prices and foreign supply. That became valuable in 2004 as surging steel demand began to push up world prices for iron ore and coal. Although these prices are reflected in Russia, owning supply moderates them.[186] The steelmakers also used their proceeds to retire their open-hearth furnaces and install continuous casters. By 2006, Magnitogorsk, which had evolved from the huge and obsolete Lenin Steel Works, had closed its last open-hearth furnace and completed its conversion to continuous casting, bringing the company to the position of second-most efficient Russian producer.[187] The steelmakers' credit position is a mixed picture. On the positive side, their ratings benefit from their low cost base and their "moderate" levels of debt. On the negative side are their continuing investment needs, the highly cyclical nature of steel demand, a high dependence on exports and a vulnerability to trade barriers, and inadequate shareholding transparency.[188] The most dramatic cost advantage is labour. At $12.50 per tonne of hot-rolled coil, labour costs are below even India's and Brazil's, and only China's is lower. In overall cost per tonne, Russia is the world's cheapest producer, followed by Mexico, India, and Brazil. The highest cost is in France.[189] Not all Asian markets are pleased to see low prices. China's steel producers' association has indicated that it will take anti-dumping action against Russian exporters if necessary.[190]

Using their high earnings, Severstal, Evraz, and Magnitogorsk have been moving upscale, and that strategy has included foreign acquisitions and new projects. North America is attractive because both the United States and Canada import steel to meet domestic demand. One reason for the large shortfall in the United States is the massive closing of bankrupt facilities, which leaves ample room for domestic suppliers. Severstal entered the North American automotive market in 2004 with its purchase of Rouge Industries, the Ford Motor Company's spun-off and bankrupt steelmaker, and in 2005 made a bid for Stelco, another major automotive supplier. In the same year Severstal joined with an investment group of former Nucor executives to found SeverCorr, a new electric-furnace facility. As was seen earlier, the mill was built in Mississippi to supply American and foreign automakers in the American South. The deal raised Severstal to the position of the fifth-largest American steelmaker.[191] In the same year Severstal acquired Gruppo Lucchini, an Italian producer of specialty and structural steels. A measure of Severstal's stature came in 2006 when it was courted as a merger partner by Arcelor in a move to foil takeover by Mittal, and when in the same year it held consultations with Europe's Corus Steel regarding a possible merger. Other steelmakers believed to be interested in Corus were Novolipetsk and Evraz, along with Brazil's CSN; the successful purchaser was India's Tata Steel.

In 2008, Severstal, which controlled 75 percent of Severcorr, bought out its partners and assumed complete ownership. Shortly afterwards, Severstal purchased the former Bethlehem Steel's large Sparrows Point mill in Maryland, which was divested by ArcelorMittal as a condition of regulatory approval of its purchase of Dofasco. Severstal then bought the outstanding shares of WCI Steel, an Ohio-based integrated producer of flat-rolled steel. It also entered a counter-offer to Essar Steel's intended purchase of West Virginia–based Esmark Inc., which operates a chain of steel service centres, following the United Steelworker's rejection of the deal. As was seen in Chapter 2, India-based Essar had purchased Algoma Steel and Minnesota Steel and intended to add Esmark to form an integrated production and distribution system in North America. In July 2008, Esmark's board approved Severstal's offer. Along with the purchase comes ownership of Esmark's other businesses, Wheeling-Pittsburgh Steel Corporation and Mountain State Carbon, a West Virginia producer of coking coal.[192] These acquisitions have enabled Severstal to draw upon the expertise of its American holdings in positioning its Russian operations for an expanding domestic automobile industry. They have also raised Severstal to the position of fourth-largest American steel producer.

Evraz initially concentrated on producing commodity-grade products and taking advantage of its Siberian locations to export to Asia. Evraz exports half of its output; its principal customers are Taiwan, Vietnam, South Korea, and the Philippines.[193] Although commodity grades are produced widely, Evraz regards the rapid growth in Asia as generating ample demand. As another part of its strategy, Evraz purchased three downstream producers – Vitkovice in the Czech Republic, Polini and Bertoli in Italy, and Oregon Steel in the United States – with the aim of supplying them with low-cost slab and selling finished products into the European and American markets. The Oregon Steel purchase works well logistically because Evraz can ship from Russia's Far East. Evraz has also engaged Oregon Steel's technical expertise to upgrade steel products at its Russian facilities. To establish a presence on the east coast, in 2008 Evraz bought Delaware-based Claymont Steel, which makes hot-rolled steel-plate products.[194] And to establish itself as a supplier to the petroleum industry, in 2008 Evraz also bought the former Ipsco Steel's minimill facilities in Saskatchewan and Alberta from Svenskt Stal AB, which had purchased Ipsco a year earlier.

For its part, Magnitogorsk in 2007 announced a plan to build a new $1 billion mill in Ohio to produce cold-rolled automotive steel. As of this writing the proposal is undergoing regulatory approval. Ohio's governor has expressed delight at the development.[195] As part of a broader expansion, Magnitogorsk in 2008 announced a joint venture to build a 2.3 mmt electric-furnace and flat-products facility in Turkey and expressed a desire to purchase Iran's 2.2 mmt Esfahan Steel Mill. Novolipetsk, the fourth steelmaker, had based its expansion on large-scale export of steel slab, and for a secure outlet purchased the Danish heavy-steel-plate producer Dansteel. In a joint partnership with Duferco, a Swiss metal trader and steel producer, Novolipetsk owns two American specialty-steel firms: Sharon Coating and Duferco Farrell. In 2008, Novolipetsk purchased John Maneely Co., the primary American producer of tubular steel products and a neighbour and customer of Duferco Farrell.

For all four Russian steelmakers, the attractions of expanding abroad reflect an additional inducement: compared to the uncertainties of doing business at home, including falling out of favour with the Kremlin, "Russian steelmen see the outside world as a kinder place."[196] Their home circumstances are indeed changeable. In July 2008, Prime Minister Vladimir Putin unexpectedly accused the Russian metal conglomerate Mechel and its owner Igor Zyuzin of selling coking coal at higher prices in Russia than abroad and of evading taxes, and threatened an investigation by prosecutors and the anti-monopoly office. The office of President Dmitry Medvedev sought to limit the frightening impact by promising a fair investigation; Putin countered by charging Mechel with price fixing, tax evasion, and harming the Russian economy, evoking memories of similar talk of economic sabotage by Stalin. One cannot but remember that the government had acquired the petroleum producer Yukos by levying crippling back taxes. Unlike oligarchs who have run afoul of the Kremlin, as did Yukos' owner Mikhail Khodorkovsky, Zyuzin had avoided political activity. In the words of Yevgeniy Yasin, head of the Higher School of Economics, "this is yet another proof that business and power are not equal."[197] It was speculated that Putin had acted on complaints from Mechel's competitors and that he was preparing the way for his friends to take over the company.[198] The implication for Russian firms is that all are potentially vulnerable.

China

Steel was a development priority of the Maoist government, whose strategy was to build small and localized installations. The result was an atomized industry of 1,042 mills, 95 percent of which operated at a loss. As recently as the early 1990s, only 30 percent of China's steel producers were using continuous casting, compared to a world average of 66 percent.[199] Lacking the advantages of modern technology and scale economies, the average Chinese steelworker in 1999 produced 41 tonnes of steel compared to the average steelworker at South Korea's POSCO, who produced 1,362 tonnes. In an effort to duplicate the examples of Japan and South Korea, the Chinese government in the 1980s began efforts to capture economies of scale, allocating much of the steel sector's total investment to one producer, Baoshan Iron and Steel Works – known now as Baosteel – to construct a very large complex near Shanghai. The government's strategic plan called for a greenfield installation using the latest technology and geared for high output. POSCO and Nippon Steel were used as models, the project was sited at a seaport, and Nippon Steel was given the design contract.[200] Much of Baosteel's technology was acquired from Germany, Japan, and South Korea in purchases and joint ventures; for China's much larger steel sector as a whole, almost half the technology was imported. With it came very rapid learning-curve effects.[201] Baosteel expanded quickly and in 2003 became the world's sixth-largest producer.

That performance would lead one to expect that China's steel industry is concentrating among a few giant producers. Indeed, among the world's top twenty steel firms, six are Chinese.[202] At the same time, however, Chinese steel producers have proliferated, rising from 1,478 in 1988 to 4,992 in 2004. The industry's level of concentration has actually fallen, with the top four accounting for one-third of total production in 1988 but only one-fifth in 2007. By comparison, in South Korea POSCO and Hyundai produce 78 percent of the country's steel; and in Japan, Nippon Steel, JFE Holdings, Sumitomo Steel, and Kobe Steel produce 76 percent. Only fifteen of China's mills have annual production over 5 mmt and a mere five have annual production over 8 mmt. If minimum efficient scale – the volume at which cost per unit is lowest – is reached at 5 mmt, only three-tenths of 1 percent of the producers have achieved it. The figure is even lower if minimum efficient scale occurs at 8 mmt.[203] Large though Baosteel is, it still produces only 6 percent of China's total. POSCO, in comparison, produces 60 percent of South Korea's.[204] Dispersed among many smaller producers, most of China's steel industry is not up to the levels of efficiency of the world leaders.

A recent study finds the reason for this in the Chinese government's decentralized administrative structure, which allows local jurisdictions to sponsor new local steel producers and to shield inefficient ones from closure. The government has encouraged the large producers to absorb smaller ones, but success has been limited to ones close by. Barriers between jurisdictions discourage larger-scale combinations. The study concludes that the result represents a failure of the Chinese government to duplicate the Japanese and South Korean models of large-scale steel development. A profusion of mills puts Chinese steelmaking out of step with the consolidating trend among the world's major producers.[205]

China's steel output has been weighted toward long products; this meets construction demand for structural steel and suits the current level of mill technology. Supplies of flat products, which include coil and sheet for manufacturing, have been augmented by imports; here, a priority is to expand production. A principal objective is to supply 80 percent of the steel used by China's automobile industry; to this end, a cold-rolled sheet mill has been built in Shanghai for plants there. The automotive sector has attracted joint ventures with both carmakers and steel producers.

Chinese iron ore is low grade, with about half the iron content of Brazilian and Australian ores. Dependence on imports prompted the joint venture just seen with Baosteel's largest supplier, Vale, to produce slab in Brazil for export to China. Baosteel has another joint venture in iron-ore production with Rio Tinto in Australia.[206] As Chinese steel production expands beyond domestic needs, the outlook over the longer term is for exporting. China, however, may not have strong export advantages. Iron ore is increasingly expensive to import, and domestic energy is in short supply, necessitating imports of coal. As will be seen shortly, labour-cost advantages are not impressive. An underdeveloped rail system imposes costs and delays on supplying domestic coal, and underdeveloped port facilities impede iron-ore delivery and finished-steel shipments. Both of these infrastructure limitations are expensive to remedy. One way of sidestepping these locational disadvantages is to move export production abroad. If production in South America expands, particularly to commodity-steel products, a nearby and wealthy market is the United States.[207]

Between 1995 and 2000, China's steel consumption increased at an average rate of 2.6 percent annually. In 2001, annual consumption growth shot up by 25 percent. In 2002, 2003, and the first quarter of 2004, demand was increasing by 26 percent *per month*, falling to 5 percent per month as the Chinese government sought to slow the economy.[208] Demand outpaced production, and China's steel imports ballooned by 48 percent between 2002 and 2003 alone. By comparison, the annual increase of world steel consumption in the same period was 2.1 percent.[209] In the first half of 2005, China's steel consumption amounted to one-third of world production, and demand was still increasing at an annual rate of 20 percent.[210] This prodigious appetite made China the top export market for the world's steel producers and changed world market conditions from excess capacity to tight supply. When world steel prices rose in 2004 to $650 per tonne for hot-rolled coil from a low of $250 in 2001, they lifted steelmakers to crests of profitability.[211]

In response Chinese capacity expanded, and by late 2005 steel production was increasing some 10 percent faster than domestic demand. Baosteel is not the only centre of growth. The Chinese government approved a $2.5 billion expansion for Maanshan Iron and Steel Company (now the Magang Group, ranked as the world's eighteenth-largest producer), and another $2 billion expansion for Taiyuan Iron and Steel (ranked thirty-first).[212] A rapid transition ensued between 2004 and 2006 when China changed, as seen at the beginning of the chapter, from a large net steel importer to a large net exporter.[213] Owing to the demand created by continuing growth in China, international steel prices have risen from $650 per tonne for hot-rolled coil in January 2005 to $1,093 as of August 2008.[214] That change was particularly significant given the slowing economies of the United States and Europe.

Although the rate of growth had been expected to moderate from the spectacular levels of the previous five years – 5 percent annually compared to the previous 10 percent – output actually grew by 15.7 percent between 2006 and 2007. Another 70 mmt of steel capacity was under construction in 2006 and an additional 80 mmt was being planned, amounting altogether to nearly half of current world consumption. As well, Chinese producers increasingly regard export markets as a natural field of expansion.[215] Previous episodes of overcapacity in the 1970s, 1980s, and 1990s caused great distress for North American, European, and Japanese steelmakers, who closed facilities, rationalized product lines, merged with one another, and sought import protection. The most vulnerable went bankrupt.

As *The Wall Street Journal* has observed, China's change from importer to exporter shows "how quickly global commodity flows can shift direction."[216] One consequence has been heightened uncertainty about world prices and supply. Given the size of China's economy and the rapid growth of its steel capacity, mismatches between production and use could strongly affect the world market and occur quickly. One prospect is that China will unload its excess, thereby sinking world prices. A second prospect is that the flow will reverse again and create another episode of tight supplies and rocketing prices.[217] The size and suddenness of such shifts add extra risk and uncertainty to the industry's usual volatility – something that has alarmed even the healthiest major producers, including POSCO and ArcelorMittal.[218] As will be seen in Chapter 5, the prospect of regulating those swings helps make consolidation a highly attractive option. Required among producers is a shared preference for stability.

Labour, Technology, and Financing

One might expect labour costs to be a developing country's advantage in producing steel. Encouraging that view is the product-cycle theory of industrial diffusion. New products and technologies are developed first in the industrial countries. As that industrial sector matures and as the technology proceeds along its lifespan, both industry and technology may migrate to newly industrializing countries, where cheap labour costs offset the accumulating disadvantage of obsolescence. Those same labour costs confer a considerable export advantage in trade with industrial countries that produce the same commodities.[219] Required is a mature, standardized, and portable technology. The human capital required is more than developing countries can supply, but being relatively modest, it is available in countries that are industrializing.[220] Do those countries enjoy a labour advantage in producing steel?

As was seen in Chapter 2, American integrated steel producers in the 1960s were overburdened with expensive labour forces and insufficiently modernized plant, and those disadvantages became a strong incentive to modernize in the 1970s and 1980s. Canadian producers, as was also seen, were better able to modernize because their strategy of concentrating on selected and profitable market sectors provided better resources. Both at the time regarded the possibilities for additional labour substitution as limited. Even Japanese mills, then seen as leaders in efficient steel production, still had large labour requirements. According to an assessment from the early 1980s, "the optimal new plant in Japan and the United States would look remarkably similar according to steel engineers ... Many of the Japanese plants require truly mammoth work forces."[221] The implication was clear: labour costs would shift the future of steel production to newly industrializing countries.

A standard economic theory supported that view. According to the Stolper-Samuelson theorem, trade raises the cost of a country's scarce factor.[222] That should make the steel industry a poor prospect in economies where labour is scarce and expensive and a good prospect in countries where it is abundant and cheap. Falling transportation costs for both raw materials and final products should augment these shifting advantages, as should locally available raw materials. In established steel-producing regions, these views justified a sombre view of the industry's future.[223]

That was the outlook then. In the meantime, high labour costs in the industrial countries have supported advances in technology. As was seen in Chapter 2, the technology of integrated steel production lends itself readily to automated and continuous processes as well as to dramatic reductions in labour content. In Canada, a survey of the primary-metals sector, which includes the steel industry, found labour productivity in that sector in 1995 to be almost twice that of the average for the business sector of the economy, and exceeded significantly only by the highly capital-intensive electrical-power-generation sector.[224] Specialization contributes savings of its own. Besides producing economies of scale, it delivers managerial efficiencies. These were seen in Chapter 2 with computerized production planning, scheduling, and communication with customers. In addition, production in large, specialized units focuses expertise, shortens the learning time for new managers, and speeds the adoption of new equipment and processes. In a capital-intensive industry such as steel, these kinds of savings "have a long-term, perhaps a permanent effect on profitability."[225] To these advantages of volume and scale can be added bulk-purchasing power for raw materials and energy. Although these are significant costs for dependent producers, firm size and market share are still important negotiating assets.

The technology that delivers these savings is standardized and for sale and reduces labour content for whomever uses it. For steel producers in low-wage countries, this eliminates an advantage. In the words of steel industry analyst Donald Barnett, "we are getting to the point where we are squeezing virtually all the water, in terms of labour, out of the production of steel. New technologies are getting rid of direct labour, maintenance, material handling. If you're comparing, say, twenty years ago – nine manhours a ton, labour costs of $35 in the States versus $10 in third-world countries – that's a big difference. But if you're comparing one manhour per ton, it's nothing – it doesn't amount to a hill of beans."[226]

As new technology is developed and installed worldwide, the opportunities for substituting cheap labour for capital continue to dwindle. The newest technologies "involve sophisticated instrumentation and micro electronics, complex software routines and exacting process controls."[227] Given the skills required for an automated mill, steel is not a low-wage product even in industrializing countries.[228] According to the calculations of United States Steel, China's labour cost advantage is only 9 percent.[229] Advantages that narrow may easily be erased by the expense of shipping to distant markets, which points to the limits of transportation costs as a locational advantage.[230] For newly industrializing countries the implication is clear: "while a number of developing nations should be able to achieve the minimum efficient scale of steel production, they should not expect to achieve any significant cost advantages from lower wage rates." Instead, competing internationally will require "state-of-the-art technical efficiency as well as economic efficiency."[231]

There the advantage may be with established producers who have aggressively modernized. According to John Surma, the president and CEO of United States Steel, the estimated cost for China to produce a tonne of hot-rolled steel is about $382. That compares to $372 for an American integrated mill and $361 for an American minimill.[232] In Surma's words, "China's labour advantage is more than offset by its raw material disadvantage, even before adding the cost of shipping." With those costs figured in, "we are better off than they are." That does not mean that China poses no threat of import penetration. The adjustment mechanism is discounted prices: according to Surma, "[Chinese producers] would be glad to sell here at any cost, at any price just to get the product to the market." American producers, he adds, should work to keep trade-remedy laws in place.[233] Pricing strategies were discussed in detail in the previous chapter; trade remedy is one of the survival strategies taken up in Chapter 5.

Labour, Technology, and Modernization

Even in the early years of expansion, labour costs were not decisive. An early assessment of Brazil's steel industry observed that "impressive growth in steel output has been at high cost to steel users yet scarcely profitable for the producers ... high production costs, heavy financial charges and taxes, sales expenses and administrative costs more than offset Brazil's low labour costs and availability of ore." Another burden was the unnecessarily large workforces imposed by state ownership. By the mid-1980s, Brazil was losing cost advantages to Korea as POSCO rapidly adopted automated technology and led the industry in reducing the labour content of steel.[234] POSCO's advances offset Brazil's primary advantage of rich iron-ore deposits and meant that to be internationally competitive, Brazilian firms would require subsidies.[235] Worse, government price controls were keeping domestic steel prices below world market levels, resulting in losses and accumulating debt.[236]

Integrated steel's inherently high capital costs make finance even more decisive than technology: "In an industry where imitation lags have consistently fallen over time, the adoption of new technologies by steel producers has become more and more dependent on capital availability rather than technology levels."[237] In capital markets that availability is related to risk and expected returns. Checkered prospects in Brazil made for expensive financing, and heavy commercial borrowing on international markets left a burden of interest payments and high costs.[238]

The state can assume capital costs through direct investment and subsidies. The Korean government, as seen earlier, directed investment funding to the steel sector. The Chinese government, which regards Baosteel as a "top-priority strategic enterprise" requiring "long-term committed support of the state in orchestrating investment irrespective of short-term losses," has funded Baosteel both directly and "through state-owned banks."[239] According to current economic geography, as was seen earlier, new centres of production and trade arise when investment locks in previously unspecific advantages. Given the capital investments required, the emergence of the Japanese, Korean, Brazilian, and Chinese steel industries as significant global producers represents a prodigious allocation of resources. Growing domestic economies generate steel demand, providing earnings for reinvestment and enabling the new enterprises to continue on their own and, through further investment, to solidify their locked-in advantages.

The new producers' ability to advance to higher-grade products depends on their ability to finance higher-end technology. Retained earnings may not be enough. The 25 percent appreciation of the Brazilian real against the American dollar in 2005 and 2006 was one reason why, in 2006, Baosteel cancelled a large and expensive joint venture with Vale to produce steel slab for China. One industry commentator observed that a stronger Brazilian currency might consign Brazil to "remaining a producer of lower value-added raw materials due to the rising costs involved in setting up more sophisticated and energy-intensive production facilities."[240] From that perspective, financing costs are an integral part of the ability to create trade advantages.

Direct purchasing, however, is not the only way to transfer technology. Arcelor has established itself as a primary producer of steel in Brazil, both for export and for an active domestic market. In acquiring Arcelor, Mittal became the proprietor of some of the world's most advanced steelmaking technology and expertise. The new company could transfer both to its operations in Brazil. Whether it will is going to depend on how the firm sees that country's domestic market and on its Brazilian subsidiaries' role in the company's overall world production scheme. If domestic demand in Brazil and demand in the firm's world market combine favourably, the appropriate technology is available to install. Bearing the cost will be ArcelorMittal.

The same transfer can occur through joint ventures that involve a steel producer in an industrializing country trading access to its domestic market for the technology held by a partner in an industrialized country. In 2003, Baosteel, Nippon Steel, and Arcelor entered a joint venture to produce top-quality steel sheet and tailored blanks in China for local automakers, including foreign joint ventures. For Nippon and Arcelor, this meant sharing their highly sophisticated technology with Baosteel. By proceeding, Nippon and Arcelor gained immediate access to China's large and growing domestic market. In return, Baosteel gained future access to Nippon's and Arcelor's specialized, exclusive, and expensive international market. One commentator wondered whether Nippon and Arcelor fully appreciated the longer-term implications of their deal.[241]

Existing facilities can be purchased directly. One asset in the merger between Ispat International (one of Mittal Steel's two antecedent firms) and ISG Group in 2005 was ISG's business with American automakers, for whom it was the principal supplier. The same access prompted Russia's Severstal in 2004 to purchase Rouge Industries, the Ford Motor Company's spun-off steel producer, and in 2005 to make an unsuccessful bid for Stelco. Arcelor and ThyssenKrupp conducted a vigorous bidding war in 2005 and 2006 for

Dofasco, whose strong financial condition added lustre to its long-standing supply relationships with Ford and Toyota. Arcelor emerged as the purchaser; then, to counter Mittal's takeover bid, it locked Dofasco into a trust to remove an asset that Mittal could sell to help finance its takeover of Arcelor. American antitrust authorities, citing a reduction of competition, made divestiture of either Dofasco or Sparrows Point, a former Bethlehem Steel mill acquired with ISG, a condition of approving the takeover. ArcelorMittal opted to sell the Sparrows Point mill and to hold on to Dofasco. Stelco, having announced that it was for sale, was purchased by United States Steel in 2007. In all of these cases, established ties with automobile producers were key attractions – especially so in light of Rouge Industries' bankrupt and aging plant and Stelco's recent emergence from court protection.

Attractive markets are not limited to the automobile industry. The American steel market overall has tempted both industrialized and newly industrializing buyers. Beyond the market's sheer size is its ability to support steel prices well above the world average, making engaging prospects of derelict facilities. Before Wilbur Ross purchased LTV in a bankruptcy auction to form the ISG Group, interested buyers included CSN of Brazil, Baosteel of China, and Dofasco of Canada.[242] The same prospects drew foreign interest to Bethlehem Steel before its acquisition in bankruptcy by ISG.

The View from International Political Economy

To place these developments in a broader perspective, it is worth evaluating them briefly in light of key tenets of international political economy. Here, the work of Robert Gilpin provides a comprehensive and familiar summary. The points of evaluation are the diffusion of technology, the respective roles of the state and multinational corporations, and the international division of labour. Although there are points of convergence with those tenets, in important ways the steel industry differs strikingly.

The Diffusion of Technology

A standard approach to explaining the expansion of industries from their original centres to new venues in peripheral countries involves the product cycle. To restate the theory briefly, firms use their home-market advantage to develop new products and technology. As these become established and perfected, the home market becomes saturated, and the initial advantages begin to dissipate. By that time the production technology has become standardized, enabling corporations to transfer it abroad. There the products are still novel enough to create demand, and the technologies may be operated by less-skilled labour. In the final stage, obsolete home production gives way to imports. The motive at play is defensive: firms seek to retain their proprietary advantage by extending it to new markets.[243]

This basic explanation has two important elements. The first is that technology is a decisively important possession. Those who are first to develop new products and production processes are in a position to earn monopoly profits until competitors arrive. In many industries that lead time can confer long-term dominance, which creates incentives for others to develop competing products and technologies. The second element is that technology maintains core-periphery structures in the international economy, with peripheral states depending on firms in core states to transfer products and technologies to them. As users of second-hand technology, they remain relegated to making established products and must rely on their advantage of cheap labour. Technological innovation remains at the core, along with control over transfer.[244]

Five important facts about the steel industry do not square with this explanation. The first is that steelmakers in the core are not developers of new technology but purchasers of it. The innovators are separate capital-goods firms, and their products are for sale internationally. They are even available with financing plans, as was seen in Brazil. Thus current steelmaking technology is available to anyone with the resources to buy it. Technological innovation in the capital-goods firms themselves is driven by cost and environmental requirements, and the results have been increasing levels of labour, materials, and energy efficiency. Producers must be able to afford new technology, and when they experience financial difficulty – as American and European producers did in the 1960s and 1970s – their ability to modernize declines, which opens competitive opportunities elsewhere. The immediate beneficiaries were South Korea and Brazil.

The second fact is that emerging producers may perfect existing technologies themselves. All of the countries reviewed began with imported technology. As was seen in Chapter 2, Japan was the first major adopter of the new BOF technology and was a very early adopter of continuous casting. Japanese producers were soon leading the way in developing high-output and automated steel production, followed energetically by POSCO in South Korea. That gave these firms important competitive advantages over their slow-to-modernize counterparts in the United States and Europe.

According to product-cycle theory, producers in newly industrializing states, because they use handed-down technology, remain dependent on cheap labour. Transfer occurs when the technology is mature as well as standardized enough to be operated by unskilled workers, making labour costs the newly industrializing country's advantage.[245] The third divergent fact is that current steel technology has resulted in drastically reduced labour content, and whoever uses it will have low labour costs. The technology also requires skilled workers. These conditions do not fit familiar images of work and production in industrializing states. Furthermore, a small labour-cost differential shifts trade advantages in steel to other factors, such as proprietary access to iron ore. Distribution of those advantages does not follow a north-south axis.

Product-cycle theory attributes technology diffusion to corporations. A fourth divergent fact is that governments of up-and-coming states on the periphery may commit considerable resources to developing their own industries and thereby gain advantages over established producers.[246] The governments of Japan, South Korea, Brazil, and China were all heavily involved in modernizing their steel industries, although their original purpose was not to export steel and challenge incumbent producers but to escape dependence on imported steel. That was part of a broader strategy to develop manufacturing sectors and lower the cost of an intermediate good. Once those sectors matured, steel was available for export and developing or exploiting trade advantages became a motivation. That same prospect troubles established steel producers as they contemplate the rapid growth of the steel industry in China.

According to product-cycle theory, technology diffusion is a lagged process that depends on the rate of obsolescence in the economies of the centre. Additional restrictions are intellectual-property rights and trade barriers, although even without them "new types of producer goods take time to diffuse around the world."[247] The newly industrializing states' incentive to gain an advantage over established producers means that, by adopting new technology, they may advance very quickly.[248] Their ability to do this shows that technology is not as exclusively proprietary as the product-cycle theory would suggest. That divergent fact certainly fits with steelmaking. The technology is proprietary to its owners, but they make it to sell to producers. Readily shipped and installed, it gives late starters the ability to catch up with and overtake incumbents.[249]

Recent evaluations of product-cycle theory have found that it best explains the stages of overseas corporate expansion that prevailed in the two decades following the Second World War.[250] Although Japan began adopting new technology and expanding its steel industry in the 1950s, it did so as part of postwar recovery. The other industries developed later. The more complex production and trade patterns that emerged are better accounted for in international political economy's view of the international division of labour, as will be seen shortly. Of primary value, however, is the product-cycle theory's emphasis on technology as key to industrial and trade advantages. That accords strongly with what has been seen.

States and Corporations

Economic geography depicts trade advantages as arising not from optimal and complementary factor endowments but from actions that lock in an otherwise unspecific set of resources. Although multinational corporations may undertake particular investments as part of an integrated international production strategy, those firms do not set out specifically to create national trading advantages. That incentive *has* guided the industrial policies of states in the developed and developing worlds.[251] Unlike corporations, states may plan and coordinate investment in pursuit of broader purposes. Their ability to do so successfully is a subject of debate; the point is that they retain the potential advantage of comprehensively planned action. The incentive to improve their countries' position internationally gives states "an almost overwhelming incentive to intervene in their domestic economies."[252]

Doing so means offsetting those market forces as well as those industrial policies of other governments that keep production and innovation in established centres. Required is a purposeful and resourceful state. But state intervention alone is not enough to create successful new centres. In addition, many states lack the means to intervene. There are examples of effective interventions, although in Japan these were indirect, and in South Korea, Brazil, and Russia they were followed by privatization. The Brazilian and Russian steel industries, moreover, only modernized and became competitive *after* the state withdrew. That said, in all cases establishing a large and internationally viable steel industry was taken as a measure of success. The same model is currently being followed by China.

Multinational corporations decide to locate production abroad on the basis of costs, access to foreign markets, and – more recently – how well particular facilities fit into international networks of production. Investment may be horizontal, with firms producing their existing goods in local markets abroad; or it may be vertical, with firms dividing production among a set of international sites, joining them together in supply chains, and producing for world markets.[253]

When the steel industry expanded to new centres, however, the initiative came not from multinational corporations but from states. World Bank loans encouraged this activity. State involvement continued a tradition of regarding steel as a vital industry and home ownership as a prerequisite for independence and security. That accounts for world steel's long-established structure of national producers. In industrializing countries (and recovering countries, in Japan's case), escaping dependence on imported intermediate goods was an added incentive. That pattern has changed only recently now that European steelmakers have begun acquiring privatized foreign firms, although with the main acquisitions limited to Brazil, the structure of the world steel industry has remained essentially national.

The current and more open-ended stage of international acquisition had its origins away from established centres. It began in Asia when the firm that is now Mittal Steel began buying privatized steel assets and grew to become the world's second-largest producer. Its sudden and unexpected takeover bid for Arcelor in 2006 alarmed the French and Belgian governments, which viewed Mittal's play as a brusque intrusion by an outsider. Arcelor's vigorous resistance campaign emphasized – gracelessly, in the eyes of critics – that Mittal was not European in either its holdings or its culture. But in the end, Mittal's success, the two firms' size, and the new entity's holdings in seventeen countries made international consolidation an immediate prospect. Chapter 5 takes up this topic, and Chapter 6 considers future configurations. Periphery-to-centre investment flows have also occurred in the United States and Canada through Brazilian, Russian, Indian, and South Korean acquisitions and joint ventures.

Vertical integration in the steel industry occurs when firms assign mills to specialize in (a) supplying other facilities in the firm, or (b) selling directly on international markets. The process is international when those assigned mills are located in various countries. An example would be a steel firm that possesses ore supplies and advantages in unfinished steel joining with a firm that possesses an established position at the upper reaches of the market. Despite these developments, the steel industry is only now beginning to acquire multinational characteristics. More vertically integrated is the world automotive industry – steel's prime customer – although its producers are now delegating component supply to outside firms. Chapter 5 examines serving the automotive industry as a survival strategy. Chapter 6 considers the relationship between trade costs and vertical integration in a globalized steel industry.

Joint ventures are a prominent feature of modern international industry. They allow corporations to pool access to markets and to share the costs and risks of developing new technologies and products. Those costs and risks are greatest when research and development (R&D) is elaborate and expensive, when significant economies of scale are involved, and when the competing firms are large and resourceful.[254] Market access motivated Japanese joint ventures in the American Midwest in the 1980s as well as POSCO's more recent enterprise in California. Global access to the automobile industry, together with meeting that industry's relentless demands for new and sophisticated steel products, has prompted joint ventures among European and Asian steelmakers. Given the automobile industry's size and dispersion, market access is an important incentive, but the major reason is that it enables steelmakers to share costly technology and expertise. More about their efforts will be seen in Chapter 5.

The International Division of Labour

One of the touchstones of explaining trade between capital- and labour-intensive countries is the Stolper–Samuelson theorem. As was seen earlier, it states that trade lowers the returns of a country's scarce factor. That happens when the scarce factor is abundant in the trading partner's economy. Labour-intensive production in countries utilizing low-skilled workers lowers the wages of low-skilled workers in wealthy countries, where the results raise serious issues of income distribution.[255] When labour-cost advantages are significant, trade forces incumbent firms to abandon products when costs are no longer competitive or to shift supply to lower-cost producers. Trade also forces incumbent producers to reduce their own labour content by adopting current technology – a process termed "defensive innovation."[256]

These have all occurred in steel, although with important differences. First, the established American integrated producers were threatened most directly by new American minimills, which lowered costs by using a simpler and cheaper technology and eliminating the expensive process of making new steel. Cheaper imports added to the competitive pressure. That pressure was strongest at the lower end in long products and hot-rolled coil, where new supply and lower prices led the integrated producers to concede those parts of their market. On the world market, the same price and supply combination made excess capacity and high costs clear to European steelmakers.

Second, "defensive innovation" had a reverse twist. The established integrated producers had delayed modernizing for the reasons seen in Chapter 2. Meanwhile, producers in Japan, subsequently in South Korea, and later in Brazil were adopting the new and efficient BOF and continuous casting technologies. That compounded the incumbent producers' disadvantages in labour costs, and their declining profits made the adjustments – abandoning obsolete facilities and purchasing expensive new equipment – even more wrenching, leaving an industrial landscape of bankrupt producers and unemployed workers. Governments directed the salvage in Europe, while in the United States the work proceeded through buyouts and consolidation. In Canada there was some of each, with the government assisting Algoma's modernization and leaving Stelco to the bankruptcy court, albeit while granting Stelco a $150 million loan and forgiving 75 percent of it provided that it eliminated its $1.3 billion pension deficit. In all cases, recovery meant saving the most modern components and installing new technology.

But the new technology reduces labour costs all around, affecting cost differences and the relative advantages of industrializing producers. The same is true more generally: in Gilpin's words, "the need to move to low-wage areas has been greatly reduced as the share of unskilled labour in production has fallen dramatically since the 1970s."[257] At the same time, differences can be seen in the steel that is traded, with incumbent producers looking to Brazil and Russia for supplies of slab while concentrating on the upper levels of the market, where they have the advantages of technology, experience, product development, and established relationships with major customers.

Whether producers in industrializing countries can move into those sectors depends on their ability to acquire and utilize the necessary technology and compete with established suppliers in new markets. Chapter 6 considers the structural changes involved. A suggestive case is the joint venture just seen involving Baosteel, Nippon Steel, and Arcelor. Supplying the world's automotive industry is the prize.

At the same time, there is vigorous trade on the northern side of the divide. The busy bilateral steel commerce between Canada and the United States is a result of industrial integration, direct and convenient linkages, and important commercial commonalities. The fact that the EU and Japan can also export and sell in North America fits a different model of trade, in which oligopolistic producers in advanced industrial countries supply one another's markets. A partial explanation is provided by the economic geography seen earlier – that is, trade is encouraged by scale economies and inexpensive transportation. The particular case of North-North trade is also explained, at least in part, by product differentiation. That condition allows Europe and Japan to sell automobiles in North America, where buyers differentiate between Cadillacs, BMWs, and Lexuses.[258]

In steel, the picture is less clear. Although demand at the middle range of the market – particularly for stainless and alloy steels – is not specific to particular producers, the most ample and strongly differentiated market is at the top end, in the automobile industry. Those customers prefer local suppliers while also demanding a complex array of specialized products. An additional element of differentiation is the automakers' preference for long-term relationships with their steel suppliers. Instead of stimulating trade, these requirements have made local suppliers attractive acquisitions. Thus, in the 1980s the Japanese developed joint ventures in the American Midwest. Fitting the mould even more exactly have been the recent Mittal/ISG, Severstal/Rouge Industries, Arcelor/Dofasco, United States Steel/Stelco, and Essar/Algoma takeovers. In acquiring ISG, Mittal became the American automobile industry's largest steel supplier.

Brazil's Gerdau was also interested in Dofasco. That fact, in addition to Gerdau's status as North America's second-largest minimill operator and the hemisphere's largest long-products producer, reflects the advance of newly industrializing countries in industries such as steel, as well as their ability to purchase assets and access in established industrial centres. The result, in Gilpin's words, has been a "pluralistic system" of foreign direct investment. This change reflects a shift of locational advantages away from natural-factor endowments and labour costs toward presences in major markets.[259] Brazil's endowment of rich iron ore counters that proposition, but Brazilian steelmakers have also benefited from a growing domestic market and their ability to invest in markets abroad. Should Baosteel begin acquiring foreign producers, its advantage will not be natural-factor endowments but retained earnings from a vast and expanding domestic market. Presence in established and wealthy foreign markets will be its motive.

Overall, steel's international division of labour reflects a transition from simple core-periphery patterns toward a more complex order. On the one hand, the trade in steel slabs reflects such a division; on the other, growing economies in countries such as Brazil and China support production further up the product line, as shown by Arcelor's involvement in Brazil and by Baosteel's joint venture with Arcelor and Nippon Steel. Closing of the north-south divide has been made possible by efficient, standardized, and readily available technology, together with a constantly growing world demand for steel. The next stage in that progression will occur when the steel industry consolidates globally and begins allocating production on the basis of world markets.[260] Overall, theories of trade and the organization of world production hinge greatly on the particular industry in question. World steel has provided the varied results just seen. Of key importance in those results has been technology.

Conclusion

We have looked at the dispersion of steel production and the growth of world supply and the way that both are affected by scale economies and linkages. Although the weight and low unit-value of iron ore and coal favour producers that locate proximately, Japan, South Korea, and China have demonstrated that it is possible to be large and successful without having those basic commodities in domestic or local supply – although transportation costs are an important variable, as will be seen in Chapter 6. Their respective governments' sponsorship of the steel industry illustrates the proposition that such interventions may secure advantages that foster successful exports. At the same time, technology has standardized production and equalized costs, removing the kinds of advantages one would expect producers in industrializing countries to enjoy. The divide that remains is mostly qualitative: the established producers in North America, Europe, and Japan have developed strong advantages at the top of the product line, in part because they have been forced by minimills and imports to abandon products at the lower end. That has given rise to a form of outsourcing in which North American and European producers import steel slab and focus on their upscale markets.

Disturbing that agreeable situation are two prospects: one is excessive expansion, particularly in China, and episodes of severe misalignment of steel supply and demand on the world market. The other is newer producers eventually ascending the product line as well. Value-added considerations provide an ample motive. A worrisome combination would be excessive expansion *and* moving into higher-end production. Then, oversupply would not be only of commodity grades. This means that the current order is by no means settled and that fears of survival, rooted in the industry's very cost and incentive structure, are never far at bay, even in good years. Survival is the topic of the next chapter, and it begins with oversupply.

5
Survival

Established integrated steel producers face threats on two fronts: prices and market share. Selling steel depends on the demand for durable goods and construction, both of which are governed by economic cycles. The steel industry is peculiarly vulnerable to downturns, as was seen in Chapter 2, because of its high fixed costs and low variable and marginal costs. These create incentives to keep producing in the face of falling prices. When that happens, sales require discounts, and with all producers governed by the same cost structure, the process readily becomes mutually destructive. Cutting production is an option in the short run, but it depends on sufficient cash reserves. It also requires other producers to do the same, which is a problem of cooperation taken up later in this chapter. Avoiding this predicament was an important reason for the industry's history of price coordination.

In the United States, coordination was possible because there were a few large producers and all had a common interest in stability. Coordination was less necessary in Canada because the various firms in the industry concentrated on their own market segments. In Europe, state ownership in some countries and corporatist economic management in others provided a safeguard against bankruptcy as well as a broader public-interest rationale for stability. More generally, in the decades of expansion following the Second World War, the troughs in the steel market were generally shorter than the peaks, which improved the steelmakers' chances of enduring downturns. Those conditions all help explain why the industry continued without significant structural change until the 1960s.

The industry has expanded from its historical centres in Europe and North America to Asia and Latin America. As was seen in Chapter 4, transportation costs and economies of scale enable distant markets to be served, and a steadily growing world demand for steel encourages trade, with over 30 percent of world production being exported. Encouraging trade as well is standardization at the low and middle ranges of the product line. Substitutability invites discounting, and the separation of domestic and foreign markets encourages price discrimination (as was seen in Chapter 3). The industry's expansion into industrializing countries has increased capacity and potentially exportable surpluses. The increase has been most dramatic in China, whose steel industry currently produces one-third of world output. The resulting leverage is strong. In the OECD's words, "most of the developments in the steel market and particularly all price movements are now closely linked to developments taking place in China."[1] Joining exporters on the world market are large steelmakers in Russia and Ukraine.

Prices are now much more volatile. Because it is difficult to forecast steel demand, and because the optimal incentive of individual producers is to expand production or at least to continue at present levels, episodes of excess supply can occur readily.[2] At the volumes now available, imbalances of supply and demand can produce sudden and large swings in price. Between a trough in 2001 and a peak in 2008 prices for hot-rolled coil nearly quadrupled from $250 to $1,093, rising from $603 in just a year.[3] With even greater volumes of exportable steel likely to be available in coming years, future swings may be more dramatic, and periods of profit may be too short for recovery.[4] The spate of steel bankruptcies at the beginning of the decade – thirty-three in the United States and two in Canada – highlights the perils of high fixed costs and weakened finances.

From all of this we can extract two survival strategies. The first involves limiting import penetration by resorting to anti-dumping laws. Successful complaints can eliminate discount margins; unsuccessful ones can still intimidate exporters into raising prices or reducing volume. The second strategy is to regulate world output by consolidating the industry. That would reproduce internationally the domestic market co-operation that once enabled producers to act on a common interest in stability. These two strategies will be examined in turn.

Another threat is shrinking market share. Established integrated producers, having already conceded lower-end products to minimills and commodity-steel exporters, will see their position at the top end eroded as producers in industrializing countries gain the technology and expertise to produce specialized steels. Technology and expertise are transferable, as was evidenced by the deal between Arcelor, Nippon, and Baosteel (see Chapter 4). Even without direct channels such as those provided by joint ventures, trade generates knowledge. Market contact with more advanced competitors "signals to [newer producers] what they must do."[5] The established producers need customers that require sophisticated new steel products; that are large, internationally distributed, and interested in collaborative long-term relationships; and that will outsource steel components. That industry is the automotive, and fostering close and durable ties with it is a third possible strategy.

Unstable Prices

A brief review of the past twenty-five years shows how changeable steel prices can be. In the 1970s and 1980s, the American steel industry adjusted to competition by abandoning obsolete facilities and lower-end products. The resulting shortages created opportunities for minimills and foreign exporters. Producers in Europe were also rationalizing their facilities, but their pace of adjustment left some 80 million tonnes of surpluses in 1981, for which the American market became a "dumping ground." While Europe was modernizing and downsizing, Japan, South Korea, and Brazil were modernizing and expanding. Because they were operating their new high-output facilities at the necessary level of capacity utilization, Japanese and Korean producers had exportable surpluses as well. Brazilian producers, driven to export to help cover foreign debts and offset the low domestic prices caused by government controls, added output of their own.[6]

After 1992, world steel demand grew strongly. Although steel producers around the world raised output, shortages had begun to appear by the middle of 1996.[7] Then in 1997, the sudden Asian economic crisis cut demand sharply, leaving stocks of unsold inventory.[8] Particularly affected were Japan and South Korea – by then two of the world's largest steel producers. Shortly afterwards an economic crisis in Russia had the same results, leaving additional unsold surplus. Currency devaluations followed in Korea, Russia, and Latin America. These raised the costs of imported iron ore and coal and also lowered the prices of exported steel; the resulting acute shortages of foreign exchange made governments desperate to sell.[9] Demand in the United States remained high, leading in 1998 to an import surge of 33 percent and a 20 percent drop in prices – and that was on top of 1997's nearly record level of imports.[10] Figures for individual exporters were much more dramatic. In the second quarter of 1998 imports from South Korea were up by 168 percent, from Japan by 140 percent, and from Russia by 31 percent. In the third quarter imports "accelerated" as "Asian steel makers intensified their attempts to export their way out of recession."[11] Surplus world steel capacity that year was estimated at 275 mmt – fully one-third of total world output.[12] In 1999, steel imports were down 14 percent but were still 24 percent above the 1994-97 average.[13]

American steel producers sought to hold back price reductions by limiting their inventories. Although earnings suffered less than in the 1991-92 recession, lower sales volume and discounted prices did leave weak producers financially vulnerable.[14] Conditions soon worsened again. In the first two quarters of 2000, sales of domestic steel fell steadily; in the third quarter, they declined abruptly. What was unusual was not the decline in the third quarter, which by then reflected a longer-term pattern, but the decline in the second one. In the Federal Reserve Bank report's words, "even during the last recession, growth in the iron and steel industry accelerated during the second quarter. The industry has not experienced a second-quarter slowdown in growth since 1989."[15] Steel prices bottomed out in December 2001 at $250 per tonne of hot-rolled coil, almost 25 percent below the 1980-2000 price average and well below the steel producers' loss point.

The same effects were felt in Canada, with steel imports rising 30.7 percent in 1997 and 15.3 percent in 1998. Between 1996 and 1998, imports rose from 30.3 percent of the domestic market to 40.2 percent. Although Canada had been a net steel importer since 1994, in both 1997 and 1998 the trade imbalance grew significantly.[16] Canada, however, had the advantage of an export market in the United States, with which it had enjoyed a steel trade surplus since 1990. In 1997, steel exports to the United States fell by only 4.3 percent; they then rose again in 1998 by 3.8 percent. With the United States representing the destination for 93.2 percent of Canada's steel exports – which themselves accounted for 29.6 percent of the value of the steel produced – that market represented an offset for the rise in imports.[17] Unlike the Asian and Russian exports to the United States, Canada's did not surge but remained close to their pattern of about one-third of production. What changed was Canada's imports from other countries: an increasing imbalance produced a net deficit in steel trade.[18]

In the United States, several years of depressed prices and lower sales, combined with the high costs of pension and benefit plans, drove thirty-three steelmakers to declare bankruptcy between 1997 and 2002, eliminating 46 million tonnes of capacity. Among those firms were the integrated sector's largest: LTV Corporation in 2000, Bethlehem Steel in 2001, and National Steel in 2002. Together they represented 57 percent of the closed capacity. The removal of that amount of steel from the market, together with a rise in demand from the automobile industry, led prices to recover by 50 percent in 2002, although they still remained below 1998 levels.[19] In Canada, Algoma Steel declared bankruptcy in 2001, and Stelco followed in 2004. Adding to the Canadian producers' difficulties were an appreciated currency and the cost advantages that had enabled the surviving American producers to avoid bankruptcy.[20]

Unwelcome Capacity

In 2001, an OECD Discussion at High Level on the world steel industry, with thirty-seven countries and the EU Commission participating, identified two problems: (1) the "survival struggles" of major established steel producers were affecting both their ability to modernize and the welfare of workers and communities; and (2) a sharply rising "volatility in trade flows" was placing pressure on governments to intervene, reflecting a "concern that the market remains fundamentally distorted."[21] In 2002, some 232 mmt of excess capacity remained in the world's major steel-producing regions: 114 mmt in the states of the former Soviet bloc, 74 mmt in Japan, and 44 mmt in the EU.[22]

The Bush administration sought to negotiate a multilateral reduction of steel capacity, and in 2001 steel-producing members of the OECD agreed to reduce their industries' capacity by 130 mmt over the next decade. Between 2000 and 2004, capacity was reduced in the OECD area by 41.5 mmt, amounting to 6.7 percent, although much of the reduction involved obsolete facilities that would have been closed in any case.[23] Obsolete capacity has been closed in the industrializing states as well; however, in response to growing domestic demand and export opportunities, new capacity is also being built. The countries of strongest capacity growth are China, Brazil, and India.[24] As a result, in the OECD's 2004 analysis, exports from the in-dustrializing states were "expected to grow – for the sixth year in a row – and pass 70 mmt, representing 28.6 percent of world exports."[25] Indeed, by 2006 industrializing states' exports had grown to 80.4 mmt, although rising overall world steel output limited their share to 21 percent.[26] Moreover, Russia's and Ukraine's steelmakers (see Chapter 4) have turned to exporting, with their contributions representing an additional 17 percent of total exports. The size of the Russian and Ukrainian net steel surpluses (see Table 5.1) means that their turn to exporting has "dramatically affected world trade in steel products."[27] And as was also seen in Chapter 4, the four large Russian steel-makers have used profits from export sales to modernize and acquire foreign producers.

Table 5.1 shows production increases between 2000 and 2007 among the largest producers and the amounts by which they exceed domestic steel consumption, with the excess representing an exportable surplus. Chapter 4, in sketching overall world trade patterns, presented Table 4.6 to show surpluses and deficits by region. To frame the problem in specific and practical terms for the discussion at hand, Table 5.1 identifies the major exporters and importers. Included are the top eleven producers (to include eleventh-place Turkey) and the NAFTA members Mexico and Canada (ranked fifteenth and sixteenth). Note well that the IISI's production data are for crude steel while its domestic consumption data are for finished steel. Thus, the figures shown in the first two columns are for crude steel while those in column 4 are for finished steel. Although the two commod-ities emerge from different stages of production, the figures are still useful for comparing relative volumes among countries as well as their trading positions – that is, whether they produce exportable surpluses or require imports to fill deficits. Precisely because countries export some steel com-modities and import others, column 6 shows net exports (surplus) and imports (deficit) in mmt for 2006, the most recent year available. This provides a general index of comparative magnitude. Those differences and the countries involved make column 6 the key focus. In the top half of the

Table 5.1

Steel production, consumption, and net trade position

Column	1 2000 production (mmt)	2 2007 production (mmt)	3 % change	4 2007 domestic steel consumption	5 2007 surplus/(deficit) (%)	6 2006 Net exports/(imports) (mmt)
Net surplus countries						
China	126.3	489.2	287.3	408.3	16.5	32.6
Japan	106.4	120.2	13.0	80.1	33.4	30.1
Ukraine	29.0	42.8	47.6	8.0	81.3	29.1
Russia	57.5	72.4	25.9	39.9	44.9	25.6
Brazil	28.5	33.8	18.6	22.0	34.9	10.7
Germany	46.3	48.6	5.0	38.3	21.2	4.9
India	26.9	53.1	97.4	50.8	4.3	1.2
Net deficit countries						
United States	100.7	98.2	(2.5)	108.2	(10.2)	(32.6)
Italy	26.5	31.5	18.9	37.0	(17.5)	(6.9)
Canada	16.5	15.6	(5.5)	15.4	(1.3)	(4.9)
South Korea	43.1	51.5	19.5	54.8	(6.4)	(4.4)
Mexico	15.6	17.6	12.8	17.8	(1.1)	(3.3)
Turkey	14.3	25.8	80.4	23.6	8.5	(3.1)

Source: IISI, *World Steel in Figures 2000*, *World Steel in Figures 2008*, The Major Steel-Producing Countries; Apparent Steel Use 2001 to 2007; The Major Importers and Exporters of Steel, 2006 (Brussels: 2000, 2008).

table, the largest net exporters appear first in column 6; in the bottom half, the largest net importers appear first.

The patterns are striking. There are five predominant net exporters – China, Japan, Ukraine, Russia, and Brazil – which together account for 81 percent of total net exports. Among net importers, the United States accounts for almost one-third of the total. Altogether, in 2007 world steel production exceeded consumption by 135.7 mmt, representing an overall surplus of 10.1 percent.[28]

It is worth considering what actually constitutes excess capacity. A common definition is "domestic capacity minus domestic consumption." The problem with that definition, according to economists Gary Clyde Hufbauer and Ben Goodrich, is that it requires a "fantasy world" of full capacity utilization and a balanced trade in steel. In their view, excess capacity is that which is "habitually unprofitable."[29] That capacity is eventually removed by market forces. Minimills have constituted those forces domestically, foreign producers internationally. The beneficiaries are steel consumers, and blocking cheaper imports transfers unearned benefits to domestic producers. That is one reason for the economics literature's generally critical view of anti-dumping actions, as will be seen shortly. Hufbauer and Goodrich acknowledge that even if excess capacity is defined as habitual unprofitability, there is nothing to prevent closed excess capacity from being replaced by new and efficient capacity, as domestic minimills have shown. If steel demand is strong, less efficient capacity may be retained as well.[30] There is also nothing to prevent capacity being built in excess of likely demand, particularly if governments and not financial markets are backing the investment. The risk is not only price competition – the focus of much of the economic critique – but also price volatility. "Persistent overcapacity," say Hufbauer and Goodrich, "has translated into cyclically falling prices and industry losses in every business slowdown."[31]

It is also worth considering what constitutes a high level of import penetration. As a measure of openness, Hufbauer and Goodrich use iron and steel imports as a percentage of GDP.[32] Among the six countries – counting the EU 25 as a single entity – importing more than $5 billion of iron and steel in 2006, those with the lowest shares, below 1 percent of GDP, were the United States, the EU 25, Canada, and China. By that measure all four were comparatively closed. The most open were Thailand at 3 percent of GDP and South Korea at 1.5 percent.[33] At the same time, figuring steel imports as a percentage of steel consumption (columns 4 and 6 in Table 5.1) showed a rather high exposure of 31.8 percent for Canada and 30.1 percent for the United States, representing the potential for international volume and price changes to exercise considerable leverage in the two countries' steel industries and making volatility in the world steel market an important concern. The period from 2003 to 2005 – when world steel demand began to soar – illustrates the point strikingly.

In the boom year of 2004, steel shortages in China and strong demand elsewhere produced a global shortage. World prices for hot-rolled coil climbed from their low point of $250 per tonne in 2001 to $650 per tonne, and in the American market steelmakers applied surcharges.[34] The high prices attracted imports and kept prices lower than they would have been with only domestic supplies.[35] A limit on imports, however, was high ocean-freight rates, caused partly by the surge of imports to China, which drove up demand for ship capacity, and by the enormous demand in China itself.[36] Even so, the shift in both volume and price was dramatic. Canada's iron and steel imports between 2003 and 2004 fluctuated by 143.7 percent and those of the United States by 86.2 percent.[37] There was a corresponding swing in operating margins; for the world's fifty-five largest producers these spiked at an average that was double the level of the trough year of 2001.[38] Some producers did very well: Arcelor's 2004 operating profits were four times greater than 2003's.[39] The same was true of share prices, with steel appearing among the top five Canadian performers of 2004.[40]

It was a very different picture in 2005, and China was one reason for the rapid change. China's steel production had increased by 24.5 percent in 2004, exceeding both domestic and international demand, even while Asian steelmakers overall continued at full volume.[41] Steel prices in China fell by 16 percent in the fourth quarter of 2005 and by 30 percent over the previous year. Major steel producers, including Mittal, Arcelor, and ThyssenKrupp, announced production cuts, and they had good reason for concern.[42] Steel prices in developed countries tend to be higher and demand patterns more stable, attracting exports when oversupply and falling prices prevail elsewhere.[43] That promised to reverse the conditions of 2004 and to depress prices in foreign markets as well as China. That is what happened. By the summer of 2005, the world spot price for a tonne of hot-rolled coil had fallen by 39 percent, which "pretty much killed profitability across the board." Even Dofasco, one of the industry's most efficient and profitable firms, turned in third-quarter results well below expectations.[44]

These large price swings and their rapidity have raised concerns about the industry's future. The concern is volume-driven volatility. In the words of Peter Hall, deputy chief economist of Export Development Canada, "there's a shakeout coming, and it's going to affect players around the world." The world steel market, he predicted, would soon be flooded with a "tsunami" of Chinese steel. The volume involved, he said, "is something we have not seen before." Compared to the steel glut produced by the 1997 Asian economic crisis, Hall predicted, "the coming shakeout will be a lot more bruising because the forces at play – primarily China – are so much bigger than anything in the past."[45]

The magnitudes are indeed formidable. Profits for China's largest seventy-seven steel producers increased by 108 percent between 2006 and 2007 alone on continuing domestic demand.[46] Meanwhile, $72 billion was invested in new mills, with state banks providing two-thirds of the funding. In 2004, there were three steelmakers in China producing more than 10 mmt per year; there are now ten.[47] The result has been 120 mmt of extra capacity. Added to 2007's output of 489.2 mmt, that total amounts to half of world consumption. In the words of Nicholas Lardy of the Institute for International Economics, "no other country even produces 100m. tonnes of steel – and that's China's excess."[48] More favourable exchange rates may not be a buffer. An increase in the exchange rate of China's currency would raise steel prices, but it would also lower the costs of importing iron ore and coal and shipping exports.[49]

At the same time, much of China's current steel export is semi-finished products. As was seen in Chapter 4, these are low-cost supplies to mills in industrial countries, representing cheaper inputs rather than lost sales. Chapter 4 discussed the advantages of using imported slab. In terms of factor costs, China's heavy dependence on expensive iron-ore and coal imports does not create a strong export advantage against producers that have their own supplies; and labour costs, as was seen in Chapter 4, are a limited advantage. Finally, most of China's steel mills have yet to achieve economies of scale.[50] Lakshmi Mittal, the chairman of Mittal Steel, believes that China needs to bring steel production into line with demand if it hopes to maintain a profitable industry, noting that the Chinese producers are aware of the problem.[51] As a measure of his confidence that this will actually occur, he has also been one of the strongest advocates of worldwide steel consolidation and managed output.

That may be a prudent approach. Commenting on the failure of Chinese steel prices to recover after their 2005 slump, Aditya Mittal, president of Mittal Steel, pointed out that Chinese steel managers are "very production focused."[52] Figures bear that out. Although 2005 was a year of surplus inventory in China, eight of its steelmakers – among a total of ten worldwide – increased production by more than 50 percent.[53] Still more capacity is in prospect in India, where Tata Steel plans three new greenfield facilities and a capacity increase of 30 mmt by 2015-16.[54] Two factors that may limit capacity expansion in China are the government's encouraging producers to consolidate – which could result in the elimination of some duplicate facilities – and to retire high-polluting older blast furnaces, which could cut some hot-metal capacity. At the same time, as was seen in Chapter 4, Chinese regional and local governments are keen to see their steelmakers expand and prosper. How these opposing forces will play out remains to be seen, although at least moderate expansion is likely.[55] In the meantime, the prospect of export surges is causing concern. In April 2007, the EU's industry commissioner, on a visit to Beijing, urged the Chinese government to limit steelmaking capacity and grant fewer export licenses. This was on the heels of a fivefold increase in Chinese steel exports to the EU in less than two years. Of particular concern was the importance of steel production to the economies of most of the EU's member states.[56]

One source of excess capacity is the world steel industry's fragmented structure. An illustration of this is the 2006 merger of Arcelor and Mittal, the world's two largest steelmakers. Their combined 2007 output of 116.4 mmt gives them less than 9 percent of total world production. One cannot say that the world steel industry is atomized among small producers; but there is no question that it is relatively unconcentrated. The IISI's roster of the top steel-producing companies lists eighty.[57] In 2007, the top fifteen steel producers accounted for only 34 percent of total world output.[58] That large and diverse population has generated a plurality of outlooks and incentives, and these have begotten persistent overcapacity: "Individual producers have made investment decisions in isolation from their competitors, both domestic and international." As a result, investment in new facilities continues "regardless of economic viability."[59] That was the perspective from which POSCO's plans to expand, seen in Chapter 4, were viewed. "Some in the industry," reported the *Wall Street Journal* in 2005, "feel that POSCO's efforts, especially in India, could add to a coming era of overcapacity, ending nearly two years of higher steel prices worldwide. Some industry leaders, such as Dan DiMicco, chief executive officer of Nucor Corp. of Charlotte, N.C., vehemently oppose adding millions of tons of steel capacity in places like India because, he said, if demand slows in those markets, the extra steel creates oversupply for the rest of the world. 'People are more focused on building new plants rather than making existing plants more efficient,' he said."[60] International steelmakers' initial response to the global financial crisis of 2008 was to cut production. As seen in Chapter 2, that option is feasible in the short run because it reduces the variable cost of iron ore and coal – the prices of which had soared in the previous three years. Continuing for more than a few months with production cutbacks, however, requires solid financial reserves. If the recession were to stretch through to 2009 (an uncertain prospect at the time of this writing in December 2008), steelmakers may turn to exporting their unsold output. One industry analyst expects steel imports to the United States to rise from their 2008 level of 18.04 mmt to 23 mmt in 2009.[61] Such an occurance would prove that co-operation in reducing output may be limited to short-term adjustments, and that in the face of longer-term downturns producers may seek to export their way out of difficulty.

Survival through Trade Protection

Relief of harm from imports is available under three different areas of trade law: anti-dumping addresses imports that are sold below home-market prices or below the cost of production; countervail addresses unfair subsidies; and safeguard counters injurious import surges. When petitions under these laws are acted upon, tariffs are imposed to eliminate the margin deemed to be unfair or to compensate for the effects of a surge. Anti-dumping is the most frequently used remedy in Canada and the United States, although the Bush administration's special steel tariffs, imposed in 2002, were established under safeguard provisions. One survey of American trade petition cases between 1980 and 1990 found anti-dumping and countervail to be the most frequently used by a wide margin: 700 anti-dumping petitions and 400 countervail petitions, compared to 25 safeguard petitions and "only a handful" of Section 301 investigations.[62] The latter is a procedure under the Trade Act of 1974 that allows the US Trade Representative to investigate the actions of foreign governments that violate their commitments under international agreements. In Canada, of the 148 trade investigations since 1984 under the Special Import Measures Act, 126 have involved anti-dumping, and 22 have involved countervail.[63] One reason for anti-dumping's popularity is that, compared to establishing an unfair subsidy, cases are easier to make. Some economists advocate relying on safeguard provisions instead because they are less discriminatory and more likely to be limited to cases of genuine harm.[64]

For steel producers seeking a survival strategy, relying on anti-dumping protection has this limitation: it places firms at the mercy of trade tribunals. These tribunals have been criticized for making petitions too easy to win; that said, their actions are not predictable, and some accounts see the willingness of American and Canadian tribunals to protect the steel industry as waning. As a striking example, Canadian steelmakers, in a major petition filed in 2002, got no relief in what they regarded as a strong case.[65] In the United States, an analysis of International Trade Commission decision making found that models according weight to political considerations, compared to models according weight to statutory requirements, do not produce significant results. One of the evaluative standards used in the analysis – "an economically rational interpretation of the statute" – makes that finding especially notable.[66] Nor do models emphasizing lobbying power perform well.[67] Results such as these do not promise steady and comprehensive protection. To understand why, it is necessary to review the broad and often technical economics literature. Steel accounts for the biggest share of both Canadian and American trade investigations; clearly, then, protection is a familiar reflex, which warrants the topical breadth of this review.

A central theme in the literature is the difference between protecting producers and protecting competition. The two are not necessarily the same, and governments acting primarily in the latter regard may not be sympathetic. Two key issues are the true incidence of predatory pricing – the circumstance in which protection is the most justified – and the welfare costs of protection. On both topics the literature has much to say. Just as important, relief even when successfully obtained may be of little help – a topic about which the economics literature also has much to say. Although the focus here is on protection as an industry strategy, the larger questions of welfare costs and instrumental effectiveness set the context in which this strategy must be realistically considered. Not all of the economics literature is negative, however, and noting when anti-dumping may be justified is a good way to begin. The critiques follow.

Justifiable Intervention

Predatory pricing, whether in domestic or international commerce, involves discounting a good to the level where competing producers are forced from the market, allowing the predator to achieve a monopoly. As was seen in Chapter 3, anti-predation is the basis of much competition law and was the original stated purpose of Canadian and American anti-dumping statutes. That purpose has been questioned. Canada's law, enacted in 1904, had no specific provision for limiting coverage to exporters with the market power to harm domestic producers; nor did it give an explanation of why the law was needed when a set of tariffs was already available. Manufacturers did not consider the existing tariff levels to be sufficiently protective. In the United States, the Antidumping Act of 1921, notwithstanding its rubric of predation, was passed by a Congress that favoured protection and was part of the duty-raising Emergency Tariff Act of 1921.[68] That act replaced the Antidumping Law of 1916 with provisions drawn from Canada's anti-dumping law, substituting the protection of producers for the protection of competition.[69] A telling index of the laws' real intent is that they were justified as ways to defend home markets. Another is their focus on the political task of regulating trade instead of the economic task of promoting efficiency and welfare.[70] Accordingly, in the words of one assessment, the "vast majority of antidumping cases are driven by the will to protect industries that claim to be in trouble."[71]

That is not to deny the possibility of genuine abuse and injury. Jacob Viner, as was seen in Chapter 3, believed short-term dumping to be the most harmful form because it weakens and drives out otherwise viable domestic producers without providing a cheaper and dependable foreign supply. That result, in Viner's view, justifies anti-dumping measures. Several recent elaborations have been added. Short-term dumping is also undesirable if it is done to avoid the costs of adjusting output to meet changed domestic demand, and if it is done to adjust to protectionism in some export markets while seeking substitute outlets in others. As well, dumping to restore a trade position weakened by a less favourable exchange rate undoes the corrective effects of currency movements and thus warrants anti-dumping intervention.[72]

Practices such as these often affect steel. Assessing the effects of subsidies, protectionism, and informal cartels, one commentary views steel as "probably the most deeply distorted industrial market in the world economy." From that perspective, anti-dumping remedies are necessary in order to maintain a liberal world trading order.[73] That notion, applied with suitable discernment and selectivity, is not incompatible with Viner's view of justifiable anti-dumping actions. And in conditions of oligopoly and increasing returns to scale – conditions that characterize the integrated steel industry – anti-dumping may actually produce positive welfare effects if it leads domestic producers to increase production and lower prices, provided they do not take surplus profits.[74]

In the steel industry, heavy fixed costs must be spread over many units of output; as a consequence, the impact of small cutbacks is strongly magnified. Cyclical dumping – pricing exports below cost at the bottom of economic cycles – is a pertinent concern because it forces cutbacks by other producers at a time when prices are already under pressure and shifts the adjustment abroad.[75] A survey of anti-dumping and countervail cases between Canada and the United States found a strong pattern of parallel investigations, with cases in each country against the other rising and falling together. One explanation is "tit-for-tat behaviour"; another is that the two countries' producers, including steelmakers, face the same market conditions.[76] On the global level, it is immaterial where surplus capacity is cut; but from the perspective of any country that is a target, "it may be considered unfair if the foreign industry can freely export its surpluses and thus increase its chances to survive." That is especially so if the exporter enjoys a protected home market. The counter-argument is that subjecting oneself to such conditions is one of the consequences of trading.[77] That consequence may be softened by membership in a trade bloc. A survey of the pattern of trade complaints in NAFTA found Canadian and Mexican exporters to be "subjected to far fewer investigations than would be predicted by their trade volumes," and cases filed to be 15 to 20 percent less successful. Of all the countries surveyed, the country with the least chance of receiving penalties from the United States was Canada.[78]

There is conflicting research on whether steel producers actually practise cyclical dumping. An early study of steel found no evidence of it, but a more recent study found a positive correlation between price troughs and the volume of anti-dumping cases and offered it as indirect evidence.[79] Additional support was seen in a longitudinal study's finding of a relationship between cyclical drops in the price of cold-rolled steel and the initiation of anti-dumping investigations.[80] Evidence that anti-dumping actions may address genuine difficulties – although not necessarily abuses – was found in the fact that the most protected industries face a significant presence of imports in their markets, employ semi-skilled workers, and have high unemployment.

Evidence that these actions were based on objective merit was found in the downturn of steel protection in the United States in the 1990s, which suggested that, in the eyes of trade officials, the industry's recovery and rationalization after the 1970s and 1980s had reduced the need for protection.[81] Even then, policy toward the steel industry in the 1970s and 1980s was "largely non-interventionist," and the actual amount of protection was low.[82] The decisions recognized foreign competition, but they also attributed the industry's difficulty to competition among domestic producers.[83] That outcome supports the view that the American trade bureaucracy is generally opposed to protection and that its pattern of relief is deliberately "too little and too late."[84]

Later, in 1992, a spate of steel anti-dumping cases was filed during a recessionary price trough. The International Trade Commission found no injury in any of the cases involving hot-rolled steel and in about half the cases involving cold-rolled steel. One reason was that, compared to the years following the Asian economic crisis of 1997, imports in 1992 were relatively low.[85] From 1994 to 1996, the United States was fourth behind the EU, Argentina, and South Africa in the number of anti-dumping cases filed.[86] In June 2001, at the depths of the steel price trough, 159 steel products from 26 countries or regions were covered by American anti-dumping or countervailing actions; of these, 95 percent were anti-dumping duties.[87] That said, anti-dumping measures have never affected more than 0.5 percent of total American imports.[88]

In Canada, the largest number of anti-dumping and countervail cases under the Special Import Measures Act of 1984 have involved steel – 32 from a total of 148.[89] In 21 of those cases, injury was found and duties were applied. In investigations of the 156 countries named in the steel cases, 147 involved anti-dumping and 9 involved countervail. Of the 87 countries named in steel proceedings launched in the period of steel-market turmoil between 1997 and 2003, 37 received findings of injury and 8 have duties that remain in effect.[90] The others had been rescinded according to a five-year termination/review provision. Of the anti-dumping measures in place in 2000, Europe and Asia each accounted for 35 percent, the United States and Mexico for 20 percent, Latin America for 9 percent, and Africa for 1 percent. Metals, at 66 percent, were the largest category.[91] Although the investigative and evaluative methods of the Canadian and American trade tribunals are different, they return findings of injury at "roughly similar" levels.[92]

"Too little and too late" was the Canadian Steel Producers Association's view of the government's response to its February 2002 safeguard petition. In August 2002, the Canadian International Trade Tribunal found five of the nine products named to be causing serious injury; it then recommended tariff rate quotas on four of those products imported from the United States.

The steel producers opposed measures against the United States because those imports entered at higher prices than other imports and flowed in even and regular volumes, and therefore did not constitute harm. Indeed, the pattern was "exactly what one would expect in an integrated market."[93] Nor did the government favour measures against the United States; or against offshore imports, fearing a negative WTO ruling. In the association's view, the government saw the choice as between acting on the tribunal's recommendations and imposing tariff rate quotas on American steel, or doing nothing and waiting for the issue to subside.[94] It chose the latter.

Yet even when actual protection is modest, the prospect of anti-dumping action makes predatory pricing risky, and a deterrent effect has been observed merely in the presence of anti-dumping laws.[95] According to game theory, potential retaliation is necessary. Ideally, governments would prefer mutual co-operation in which no producer takes advantage of open foreign markets, but without the means to punish predatory behaviour, governments are likely to "see beggar-your-neighbour policies as rational."[96]

Subsidies – government payments to a producer to offset costs – allow exports to be discounted and are subject to the trade law of countervail. Like anti-dumping laws, countervail allows subsidy margins to be eliminated by tariffs if the subsidy is unfair under the terms of the law and is causing injury. Subsidies have indeed existed. One indication of this was the change in POSCO's pricing practices after the company was privatized in 1999, removing the influence of the government's financing and export policies.[97] Another was the EU's decision to accept a voluntary export restraint on steel exports to the United States instead of allowing anti-dumping measures against individual members. Selling surpluses in the United States was the behaviour in question. Anti-dumping tariffs on the exports of individual EU members would reflect the level of discounting, would be high enough to curtail exports, and would cause "massive" trade diversion in the barrier-free EU.[98]

Unwarranted Intrusion

Now for the critiques. One of the strongest is that predation – the target of anti-dumping laws – actually occurs quite rarely. Using the stricter standards of domestic competition law as a gauge, an unpublished OECD study found that fewer than 10 percent of the cases filed in jurisdictions that often use anti-dumping laws involved situations where "authentic predation *might* have been possible."[99] A related question is how to accurately identify abuses. According to one critique, "there appears to be no distinction between (profit maximizing) price discrimination and predatory behaviour in the trade literature." That observation raises a related one: anti-dumping punishes actions that are normal in domestic commerce. A good example is cyclical dumping. Encouraging the practice is not only surplus output but a higher demand elasticity in foreign markets, which increases the response to lower prices.[100] In domestic markets, it is "thought to be common and rational" for firms with high fixed costs and facing cyclical demand to continue production at lower prices.[101] Surplus capacity encourages cyclical dumping; thus the incidence of anti-dumping petitions varies with fixed domestic capacity and surplus foreign capacity.[102] If trade remedies tend to be applied most frequently during economic downturns, however, the question arises whether foreign competition is fair during periods of prosperity and unfair during periods of distress.[103]

Under competition law, concentration is a necessary condition for establishing predation, and oligopolistic industries are potentially areas of concern.[104] Concentration at levels sufficient for predation, however, has been found to be the exception. In a sample of 169 successful anti-dumping cases in the 1980s, only 39 involved industries characterized by "substantial" concentration.[105] And even when concentration exists, foreign entrants who displace domestic producers will not raise prices to monopoly levels if the market is contestable – that is, open to other domestic and foreign firms – because prices very far above costs would attract new competitors.[106] Note that the emphasis is on competition overall and not on individual producers. One source of contestability relates to an anti-dumping tariff's trade-diverting effect on other exporters. Their readiness to sell "suggests that a predatory strategy cannot easily succeed."[107]

In any case, predation looks to be rare.[108] In a sample of thirty Canadian cases, predatory pricing was present in none. In twenty-six of the thirty, a contestable market and alternative exporters curbed predation. And on the question of protecting domestic producers against oligopolistic exporters, twenty of the thirty cases involved a Canadian company that either possessed "oligopolistic market power" itself or was an inefficient producer.[109] Finally, resource-based industries such as steel have been found to be unlikely candidates for predation because they produce standardized commodities that are "subject to enormous competition worldwide."[110]

Price discrimination, as was seen in Chapter 3, makes it possible to charge different prices in home and export markets. Doing so, however, may actually improve welfare by allowing economies of scale to be more fully exploited and by making goods available that might not be otherwise.[111] As well, countries as a whole can benefit from trade even if the balance of advantages is one-sided and artificial.[112] Viner acknowledged this in his view of long-term dumping, in which a foreign producer sells in an export market at a discounted price. Even though that may drive domestic producers out, the country as a whole benefits from a cheaper good. Again, that view is predicated on the interests of the overall economy, not on those of individual producers. Less drastically, discounts that benefit the economy are acceptable as long as domestic producers are able to stay in business. Dumping is undesirable if it causes "greater aggregate social loss and injury to domestic import-competing suppliers than benefits to domestic buyers."[113] Likewise, protection is undesirable if it weakens "the static and dynamic forces of competition" on which a country's overall economic welfare depends.[114]

Anti-dumping duties on intermediate goods such as steel raise costs for manufacturers, who suffer if those costs cannot be passed along either to final-product makers or to consumers.[115] For the national economy there are welfare losses when exports are forgone because of increased costs.[116] Predictably, downstream users who are unable to pass along costs are political foes of protection, as will be seen presently with the opposition to the Bush administration's special steel tariffs.[117] Consumers do pay. One analysis of patterns of protection in Canada and the United States found that duties are more likely to be applied to goods for which consumers bear the extra costs.[118] Protection may also encourage capital-heavy firms to employ more labour. When labour has sufficient bargaining power, it can appropriate the rents from protection and prevent them from being passed along to consumers.[119]

Welfare gains may also be thwarted if protection attracts new firms and prevents incumbent firms from expanding production and lowering prices. Efficient allocation of productive resources is more likely when tariffs are reduced or eliminated.[120] In the same way, protection shields inefficient firms from the pressures that would force them to modernize, representing another forgone welfare improvement.[121] Monopolists, for their part, can earn larger profits.[122] As a final point about welfare effects, imports may inflict only modest levels of injury. When that is so, even low levels of protection still impose unnecessary costs. A finding of low actual injury emerged from an analysis of 105 cases between 1980 and 1988, of which 36 involved steel.[123]

A related theme is perverse outcomes. As one example, the benefits of anti-dumping protection may go to exporters. This can happen in several ways. Exporters raising their prices to avoid an anti-dumping ruling may end up increasing their revenues. Also, if the restriction is quantitative – that is, if an exporter agrees to limit volume to a stated level in order to avoid duties – the result may in fact stabilize normal market fluctuations, to the exporter's benefit. Furthermore, should demand increase, a quantitative agreement leaves the extra margin to be filled by domestic producers, who may exploit the situation by raising prices.[124] Another possible perverse result is collusion between domestic firms and exporters, who could agree to raise prices or lower volumes, producing a "collusive equilibrium in which both parties earn higher profits."[125] Although that outcome is attractive in theory, a survey of steel cases found that exporters generally chose to pay the anti-dumping duty instead, in order to preserve market share by retaining the ability to decide the portion of duty to include in prices.[126]

In addition, protective measures may be self-defeating because they motivate foreign producers to compensate for duties by raising their quality: "When a foreign firm is a low-quality producer under free trade, a price undertaking will turn it into a quality leader in the market." The result is "product diversion."[127] Protective measures also work against the long-term interests of domestic firms by inviting "tariff jumping" – that is, the establishment by foreign producers of facilities within the protecting country. Because this tactic requires considerable capital and organizational adaptiveness, it is most open to steelmakers from other advanced industrial economies, the very ones who would be the most effective competitors.[128] Duties against imported steel may also have the perverse effect of opening market share for domestic minimills, making them free riders on the integrated producers' protection. Even if the duties are only on upper-end products, protection still provides minimills the incentive to develop that technology.[129] On the other hand, if protection shifts production to the integrated producers, it is a welfare loss because minimills could have made the products more cheaply.[130] Additional free riders are foreign steel producers not covered by duties, who may also enjoy new market share.[131]

Does Protection Actually Help?

A related question is whether tariff protection actually offers much aid to producers. Here, a major instance to evaluate is the Bush administration's special steel tariffs of 2002. On June 5, 2001, the administration announced a two-part strategy for the steel industry: it would initiate negotiations at the OECD for multilateral reductions of steel capacity, and it would launch an investigation under Section 201 of the Trade Act of 1974 to determine whether imports were causing or threatening to cause serious injury to the steel industry. Section 201 differs from anti-dumping and countervail measures in that the imports in question need not be traded unfairly. Under WTO rules, the measures must be temporary and must be used to allow domestic producers to adapt. Domestic political considerations were involved. The departing Clinton administration "waited until the penultimate day" in office to recommend an investigation of injury, leaving the matter as fresh business for the incoming Bush administration.[132]

The issue itself had been pending. In 1999, the House of Representatives had passed a bill that would have imposed quotas on all steel imports. The Clinton administration feared that the bill would violate WTO rules, attract retaliation, and impose large costs on domestic steel users, and threatened a veto. The Senate defeated the bill, sparing the administration from having to offend labour in the approaching presidential election.[133] Soon after the election, during the Senate confirmation hearings of the Commerce Secretary nominee Donald Evans, steel-state senators made it clear that congressional support for fast-track authority for the new Bush administration to pursue trade agreements would be conditional on attending to problems in the steel industry.[134] In June of that year, Evans announced the Section 201 investigation. Expressing approval were both representatives of the steel industry – which had donated $2 million to the Republican Party for the 2000 elections – and the United Steel Workers of America, which indicated that the decision "could help Bush in getting the union's endorsement" in the 2004 elections.[135]

Acting on the International Trade Commission's investigation of thirty-three steel products and its finding that sixteen of them were causing substantial injury, the Bush administration on March 5, 2002, announced tariffs of 8 to 30 percent, which were to remain in place for three years. Canada and Mexico, as NAFTA members, were granted exemptions, as were developing countries. In addition, exemptions or reductions were made on appeal from exporters and domestic steel users; eventually these amounted to about one-quarter of the items on the original list. In light of the WTO's record of ruling against blanket measures such as these, a pragmatic calculation was that the tariffs would be in force only until the major steel-exporting countries organized their challenge and obtained a WTO ruling. That happened sooner than many expected. In July 2003, the WTO ruled against the tariffs; that ruling was upheld on appeal in November, opening the way for complainants to retaliate. The Bush administration repealed the tariffs on December 4, just eleven days before the EU was set to impose some $2 billion of tariffs against an array of American exports, which had been carefully selected for political and symbolic impact. By that time, there was considerable political counter-pressure from steel-using industries, which argued that the tariffs were severely raising their costs. Many of those industries were in states that would figure strongly in the upcoming 2004 elections.

The steel industry by then was on its way to recovery. Proponents of the tariffs argued that firms had been given crucial time to restructure. A recent econometric analysis, however, indicates that credit should not be given to the tariffs. Improvement came instead from lower production capacity owing to bankruptcies and reorganization, economic recovery after the 2001 recession, and import price increases from a falling dollar. The rise of demand from China also played a role, but with a six-month lag. There was "little evidence" that the tariffs themselves had a beneficial effect. That finding is in keeping with other economic analyses over the years, which have found the steel industry to be "relatively immune" from protection.[136]

The considerations just seen, both theoretical and practical, raise the question of whether protection is a good long-term strategy. If massive interventions such as the Bush tariffs have little effect, if the costs and counter-effects of protection are significant, and if there are relatively balanced groups of winners and losers, the whole strategy seems dubious in both effectiveness and predictability. A paring back of anti-dumping measures was on the agenda of the Doha Round of the WTO; however, steelmakers advocated leaving them in place as a needed defence against discounted imports. If they do remain, uncertain and episodic security is the prospect.

Surviving by Merging: Global Consolidation

Mittal Steel's surprise merger with Arcelor in spring 2006 was the steel industry's equivalent of the energy industry's union of Exxon and Mobil.[137] The new company, ArcelorMittal, has business operations in sixty countries and continues to expand its holdings. The company's 2007 annual report summarizes a busy program of acquisition, which in 2007 amounted to $12.3 billion.[138] The unprecedented scope and pace of the company's activities has made steel consolidation a pressing issue, especially since further unions are expected. As one indicator of sudden change, prior to 2004 there had been little discussion in the financial and trade press of international mergers among the industry's top firms. As one indicator of pace, Mittal's takeover occurred relatively quickly even though it was fiercely resisted by Arcelor. Other large-scale mergers could also occur quickly, either to match ArcelorMittal's advantages or to defend against being absorbed by it. So it can be said the world steel industry's basic structure is today subject to rapid and far-reaching change. Enough information is now available in the financial and trade press to consider whether consolidation is a useful survival strategy and why it might be attractive. One of the primary motives for consolidating is to reduce the number of independent centres of decision making in order to facilitate coordination and limit volatility.

Consolidation involves combining firms under a single management. There are several prospective advantages to doing so. One is that the surviving elements can be reorganized into an efficient entity by eliminating capacity. A closely related advantage is that it enables the new firm to improve economies of scale by melding facilities and unifying management. A third advantage is that it allows the achievement of synergies – a key reason why Mittal Steel acquired Arcelor. That union has blended Mittal's efficiencies and market position in the low and middle sectors of the product line with Arcelor's development capabilities and market position in the high sector. Another synergy is that the Mittal/Arcelor merger has combined ownership of iron-ore supplies with ownership of top-end products and an established presence in wealthy industrial markets. That was the rationale behind Severstal's and Corus Steel's preliminary interest in a merger. A fourth motive is that merging enables firms to acquire heft for dealing with powerful counterparts. Even after the Mittal/Arcelor merger, the top fifteen steelmakers, as was seen earlier, control only 34 percent of the market; however, they also deal with an oligopoly that controls 75 percent of the export iron-ore market and with an automobile industry whose top three members control 50 percent of that market.[139] A final motive is defensive: to boost scale economies, market share, and revenues in order to survive in competition with much larger producers. All of these considerations have been at play in the surge of interest in consolidation evoked by the Mittal/Arcelor merger. For the most successful producers, profits from the record year of 2004 provided the means of acquisition.

There has already been considerable consolidation at the national level. In the United States, mergers as well as purchases of bankrupt assets have reduced the number of integrated producers and increased their market share. The change has been rapid and impressive. In 2002, seven producers claimed half of the American market; in 2003, that share was claimed by three – United States Steel, Nucor, and Mittal Steel. The Ispat/ISG merger in 2004, which formed Mittal Steel, united five of the seven largest steelmakers that existed in 2001: LTV, Bethlehem, Weirton, Acme, and Ispat Inland (Mittal's integrated American producer, acquired in 1998). United States Steel had acquired National Steel in 2003.

In Canada there has been no concentration among the integrated producers. One reason why is that all three are viable. That may have been a doubtful claim when Stelco declared bankruptcy in 2004, but record profits that year enabled it to emerge from bankruptcy twenty-six months later. Dofasco did acquire Algoma Steel in 1986, but then sold it in 1992. After Algoma emerged from bankruptcy in 2004, its profits and cash reserves recovered to the point that major shareholders demanded higher dividends, and the company became an attractive acquisition for Essar Global Ltd. In Europe, the major consolidations have been the 1999 merger between privatized British Steel and the Dutch producer Hoogovens, which resulted in Corus Steel, and the 2002 merger of the Spanish producer Aceralia, the Luxembourg producer Arbed, and the French producer Usinor, which produced Arcelor. In Japan, Kawasaki Steel and NKK Steel combined in 2002 to form JFE Holdings. In Latin America, Arcelor's acquisitions have made it Brazil's largest producer. By the time it merged with Arcelor, Mittal had assembled a global firm from producers in an eventually long list of locales: Indonesia, Trinidad, Mexico, Canada, Germany, the United States, France, Algeria, South Africa, Kazakhstan, Romania, the Czech Republic, Poland, and Ireland.

At the global level, however, the steel industry is still not concentrated. Before the Mittal/Arcelor merger, the ten largest producers each accounted for only 2 to 4 percent of world production – the same pattern that had prevailed in the 1970s. Growth in steel demand was accommodated by increased production that saw the nineteen largest firms double their output over that of 1990. Some of that increase did result from consolidation; even so, the top ten steelmakers still produced only one-third of world output.[140] From that perspective, ArcelorMittal's combined market share, while still only about 10 percent of world capacity, is still unprecedented. "For the first time in the steel business [the merger creates] what many analysts describe as a truly global player."[141]

Compared to Mittal Steel after its merger with ISG and especially after its merger with Arcelor, American and Canadian producers are second- and third-tier producers and have a small presence internationally. That said, their established presence in North America makes them potentially attractive acquisitions for foreign producers. Access to local markets motivated Arcelor's purchase of Dofasco in 2006; other candidates of similar size are the American producers AK Steel and Steel Dynamics. United States Steel and Nucor are seen as not large enough to survive on their own in a consolidated industry and are potential quarry.[142] For smaller producers such as pre-acquisition Dofasco, a limit is resources to finance research and technology to remain competitive at the top end of the product line.[143] They risk losing customers because they cannot keep pace with innovations by larger producers. So far, steelmakers such as Arcelor and Nippon, which have been at the forefront of R&D and to whom others have looked for practices to emulate, have not installed their own operations in North America. If they do, they will bring their leadership advantages with them.[144]

A primary incentive for industry consolidation is that it enables firms to coordinate production to reflect demand and thereby avoid or moderate the industry's volatility. "Extreme fragmentation," in the words of Severstal's president Alexi Mordashov, "has been an endemic problem. Historically companies have not had significant power to stabilize prices. This has led to significant shakeouts ... The only way to avoid the repeat of such recurrences is through consolidation of the industry into a handful of companies."[145] One advantage of size is that it enables a firm to affect overall supply and price. To offset excess supply and falling prices in 2005, Mittal Steel cut production by 1 mmt in both the second and third quarters. Arcelor and ThyssenKrupp announced production cuts as well. Mittal's chairman, Lakshmi Mittal, said that the move would slow the buildup of inventory and restore equilibrium in the "global supply and demand equation."[146] In the view of Mittal Steel president Aditya Mittal, the decisions were a milestone. Addressing a steel conference in London in November 2005, he declared that "for the first time in the history of the steel industry we have been able to sustain value in an over-supply environment by temporarily cutting production. This is a major step change in the way the industry is behaving largely due to consolidation."[147] The strategy is not limited to steel. Former American treasury secretary Paul O'Neill, when he was CEO of Alcoa Aluminum, supported an aluminum cartel that has been able to coordinate increases in world prices.[148] Responding to the global financial crisis of 2008, Lakshmi Mittal, the principal owner of ArcelorMittal, announced in November that the firm would cut production for the remainder of the year by up to 35 percent in recognition of severe market conditions. Such unilateral moves may tempt others to retain their levels of production, but Mittal was confident that his firm's position as the industry's largest firm would enable its reduced production to influence the others' pricing decisions.[149] Some major producers also cut their production before Mittal announced his move, so that his firm's informal position as the industry's price leader is not yet clearly established.

Size is an advantage. One commentary on the Mittal/Arcelor merger stated that "larger companies, particularly in a cyclical industry – have a greater ability than smaller ones to moderate their production to fit with demand patterns to keep prices and profitability high."[150] Another advantage of consolidation is that it enables a firm to be present in multiple markets and thereby exercise multiple points of influence. Other commentators have expressed doubt that an 11 percent world market share will translate into pricing power but agree that the prospect will spur further consolidation, which in turn will lead to a beneficial "element of price discipline."[151] These developments have not been welcomed by all in the industry. The president of an American minimill contends that "the big players are after regional control. They want to buy capacity and just put it into the market as they see fit."[152]

Achieving economies of scale is another attraction. These economies are important in steel because they allow high fixed costs to be spread over a larger output. This advantage has cash value. When Ispat International merged with ISG to form Mittal Steel it paid a premium price, and prospective scale economies were an important consideration. The merger combined ownership of three adjacent facilities on the south shore of Lake Michigan, and their proximity allowed management to be integrated, supply to be organized into a single stream, and plant output to be specialized.[153]

That accords with a key consolidation strategy: "Re-organize rolling mills and processing lines to rationalize the probably diverse, but overlapping, mix of product size and grade ranges that a portfolio of acquired plants is capable of handling."[154] When some of the assets in a merger are redundant or obsolete, the deal represents an opportunity for "selective modernization."[155] The decisions related to that process were seen in Chapter 2. Overall, according to one scholar of the industry, "we foresee fewer steel firms and fewer steel plants. Logic suggests scale economies and the need to make a wide range of products for powerful customers will result in greater concentration of production. Steelmakers have to pursue scale and scope – both size and product range, or find a niche."[156] There are also defensive reasons to achieve scale economies. An important one is to lower production costs to withstand bouts of falling steel prices or an appreciated currency.[157] That requires building up enough reserve cash or credit during periods of high prices to survive periods of low ones. This task is hardest for smaller and high-cost producers.[158]

A third attraction of consolidation is that it enables a firm to gain the collective power to deal with consolidated suppliers.[159] As was seen in Chapter 2, the world's iron-ore industry is a powerful oligopoly, one that controls some 75 percent of the world's ore supply and that exercises strong pricing power. It was able to impose a 71.5 percent price increase in 2005, an additional 20 percent increase for 2006, and a 65 percent increase in 2008. As was also seen, the industry price is set by the deal struck by the oligopoly and Japanese steel producers. The fact that Japanese steelmakers – among the largest in the industry – were forced to accept the increase is itself impressive. An even stronger indicator of the ore producers' pricing power, however, was their ability to force the Chinese government's negotiators – who were under strict instructions not to accept more than a 10 percent increase – to settle for 19 percent (nearly the full amount) in 2006 and to accept a large transportation premium for Australian ore in 2008. Even the world's largest iron-ore importer was unable to use its purchasing power defensively.[160]

Outcomes such as those make independent ore supplies an important advantage. One benefit of Ispat's merger with ISG was that it ensured access to North American ore supplies, an advantage shared by United States Steel, Algoma, Stelco, and Dofasco. Along with other ore acquisitions and its merger with Arcelor, that merger has made Mittal the world's largest iron-ore producer after Vale, Rio Tinto, BHP Billiton, and Ukraine's Privat Intertrading.[161] A study by Integer Research found that the steel industry's most profitable companies had close access to iron ore and coal. Of the world's ten most profitable producers, seven are located in Russia, Ukraine, and Brazil. The three most profitable are Russia's Novolipetsk and Brazil's Usiminas and CSN.[162] Mittal does purchase roughly 50 percent of its iron ore from the other large producers, but its own holdings do confer it some bargaining power.[163] Even so, in 2007 ArcelorMittal purchased a Brazilian iron-ore mine. The experience of the Chinese and Japanese in their 2005 and 2006 ore price negotiations showed that sheer size may not be enough against a powerful oligopoly; that said, a world steel industry of perhaps ten producers would be better able to negotiate as a united counterpart.

A fourth advantage of consolidation is that firms thereby acquire facilities to supply raw steel for finishing into high-margin products. As was seen in Chapter 4, producers in Brazil and elsewhere sell slab steel to mills in North America, Europe, and Japan, and producers such as Russia's Severstal and Brazil's CSN have the advantage of their own iron-ore supplies. Integration with downstream producers located in large industrial markets combines the advantages of both operations under a single ownership. That combination motivated Mittal to acquire Ukraine's Krivorzhstal and Arcelor to acquire Brazil's CST. The investment can also flow in the reverse direction, as when Brazil's CSN purchased Heartland Steel. Consolidation brings complementary assets under a single ownership, allowing metal commodities to be circulated internationally within the structure of the given firm.

A final advantage is that consolidation offers a firm access to the merger partner's markets. Many targets for acquisition are in the mature industrial countries. As was seen in Chapter 4, in 2004, the Russian steelmaker Severstal acquired Rouge Industries, formerly the steel-producing branch of Ford Motor Company and that automaker's largest steel supplier. As Severstal North America, the firm intends to carve out a steel-supplying role for all of North America's major automobile producers.[164] In 2005, Severstal submitted a bid to buy Stelco, and Arcelor submitted a friendly offer for Dofasco. Arcelor's intention was to acquire Dofasco's long-standing supply relationships with Ford, Honda, and Toyota and to gain entry into the wider North American market.

The same prospect had made Dofasco of interest to the Brazilian steel producers Usiminas and Gerdau, which also wanted direct access to North America's automobile industry.[165] Shortly after Arcelor made its offer, Dofasco accepted a higher bid from Germany's ThyssenKrupp, who sought the same access. A bidding war ensued, with Arcelor emerging in January 2006 as the buyer. In the meantime, Algoma's situation had improved enough to make it an attractive takeover, and it was purchased in 2007 by Mumbai-based Essar Global, India's largest flat-steel exporter. These events can be seen as aligning steel ownership and production with customer preferences for local suppliers – preferences that make serving the automotive industry highly attractive, as will be seen shortly. Noteworthy about these and similar deals is that some come from producers in formerly peripheral areas – notably Russia, Brazil, and India – and are directed at assets in the mature industrial countries.

The possibility of access to new markets has also sparked interest in mergers with Chinese steelmakers. In 2005, Mittal purchased a 37 percent stake in Hunan Valin Steel Tube and Wire, and Arcelor purchased a 38.4 percent stake in Laiwu Steel Group. There have not, however, been major foreign acquisitions. The Chinese government has declared that no foreign steel producer will acquire majority shares and that the industry will be rationalized to ten major producers. More interesting, it asserts that Baosteel, instead being a takeover target, should be a global consolidator. Both the Chinese government and outside observers such as Aditya Mittal have advocated mergers among Chinese producers themselves, and many expect that rationalization and consolidation will occur as the industry matures. These intentions notwithstanding, as was seen in Chapter 4, smaller steelmakers in China have actually proliferated. For their part, neither United States Steel nor Nucor believes that China is a good place to invest.[166]

A spate of steel mergers may be one effect of the 2008 global financial crisis. By December 2008, production cutbacks and collapsed prices had weakened a number of firms, but others possessed ample resources for take-overs. Among the potential buyers, according to analysts of the industry, were ThyssenKrupp and US Steel, as well as some large but unnamed Chinese steelmakers. Candidates for acquisition were said to include Corus, the Italian steelmaker Rive, and Swenden's Svenskt Stal AB. That development is prospective as of this writing, but if the mergers do proceed they would represent another step in the industry's concentration.[167]

Some of the steel industry's most valuable customers are themselves consolidated. The automobile producers are the integrated steelmakers' single largest customer, and in the past their buying power enabled them to bargain for low prices. That ability depended on the two sides' relative strengths during any particular negotiation. During the steel shortage of 2004, automakers had to accept large price increases; in previous years, when steel producers were coping with low prices and financial distress, those automakers had the upper hand.[168] Large mergers alter the relationship. "Some customers worry that the new breed of larger steel companies will have greater power to dictate price, quality, and availability of steel in an environment in which steel prices are already higher than in recent years."[169] As steel prices continued to climb into 2008, the automakers' costs and pricing ability came under further pressure; even so, they remain committed to steel as their primary ingredient.[170] Other important steel buyers, particularly small manufacturers, construction firms, and steel service centres, are not concentrated and are much more likely to be in the position of price takers.

Customer reaction to the Mittal/Arcelor merger, accordingly, was mixed. A source of concern was the ability of fewer and more powerful steel producers to limit customer bargaining power. In 2004, Ispat Inland, Mittal's American unit before the ISG merger, imposed a price hike on Whirlpool Corporation, one of the largest appliance manufacturers. Ispat, which "already has a hard-nosed reputation," forced Whirlpool to raise prices by 10 percent. In the words of Whirlpool's chairman and CEO, "I don't think any company can overcome that magnitude of raw-material price increases with productivity."[171] Other industries are vulnerable as well. In 2004 and 2005, five major manufacturers of automobile components went bankrupt, and a study commissioned by their trade association, the Motor and Equipment Manufacturers Association, found them caught between the steelmakers' high prices and the automobile producers' refusal to pay more for components.[172] Feared more generally is a long-term shift in bargaining power from automakers to steelmakers.

One response has been to encourage other global steelmakers to enter the North American market to provide more competition, and senior officials of both General Motors and Ford have acknowledged that they have conferred with major foreign producers. In the words of a Ford Motor Company spokesperson, "We'd want to be assured that a possible significant increase in market concentration would somehow improve competition rather than reduce it at the expense of immediate customers."[173] One attraction for foreign producers is market potential. GM's annual steel purchases amount to $3 billion.[174]

An alternative view is that steel customers benefit from a stable and healthy industry. Consolidation can be worthwhile if it makes the industry better able to match inventory to demand and diminish the industry's characteristic and cyclical instability. Uncertain of the benefits of consolidation are the buyers of highly specialized products of the sort that depend on a small number of producers.[175] From the steelmakers' perspective, there are benefits to be had from exercising moderation when dealing with customers. The CEO of Mittal Steel's American division has stated that "there is no market we want to lose." There is also the option of turning to imports, although tight output and price coordination in a consolidated world steel industry would limit price ranges – perhaps to the extent of eliminating imports as an option.[176] Moreover, to use imports, particularly for sheet steel and other products that must be coordinated in closely managed supply chains, would be to forgo the predictability of local supply.

In the meantime, in one steel executive's words, "we often describe ourselves as kind of being caught between a highly consolidated supply industry and an increased consolidated customer base and we are still very fragmented."[177] The steel industry, in the view of a coal executive, is the "ham in the sandwich" between powerful suppliers and customers. Without consolidation to achieve commensurate bargaining power the industry's "customers and suppliers will continue to exert price outcomes."[178]

The response of other major steelmakers to the Mittal/Arcelor merger was attentive. Producers were keenly aware that the deal was a hostile takeover and that similar moves might consolidate the industry quickly and massively. As a defensive measure, Tata Sons Limited, owners of 20 percent of India's Tata Steel, increased its holdings in the company by 10 percent.[179] In Japan, Nippon Steel, Sumitomo Metal Industries, and Kobe Steel, believing that the world steel industry was "undergoing consolidation on a global scale," agreed to co-operate to thwart hostile takeovers.[180] Nippon Steel, for its part, has adopted an American-style poison-pill defence against takeovers.[181] The reaction of the president of JFE Steel – itself the result of a recent merger – has been that the company is open to further integration and that the Japanese steel industry still has many minimills and smaller specialty-steel producers that are takeover candidates. At the same time, he has noted that the Arcelor/Mittal merger reminds managers always to "be aware of the possibility of becoming targets of hostile takeover attempts."[182]

Various future configurations suggest themselves. Before the Mittal/Arcelor merger, the most dramatic forecast was for two major global steel cartels: Arcelor-POSCO-Nippon and ThyssenKrupp-JFE.[183] Another forecast, looking beyond 2010, saw the industry as cartelized to the degree of coordinating common prices electronically: "The computers of all companies link up. They take over the price negotiating process with customers in both the spot and one-year contract markets. The result is an 'administered' pricing system never imagined in economic textbooks. Prices rise when volume declines because fixed costs are higher."[184] Although such coordination raises the prospect of abusive pricing, governments may have reason to be indulgent. Home-owned steel industries are no longer regarded as necessary for national security, and governments in all of the mature industrial countries have had to cope with steel-sector rationalizations and layoffs. Willing foreign buyers have revived dormant or doubtful assets such as Bethlehem Steel, and governments have been strongly aware of the wages of oversupply and price volatility and of the burdens of picking up pension benefits. Whether this indulgent view could change is a topic of Chapter 6.

Consolidation itself does not solve the problem of excess capacity. Commenting on new alliances between European and Japanese steelmakers (which will be discussed in the following section), an OECD study observed that "in the long run ... the steel industry will have to close a substantial number of production lines to cope with its huge capacity surplus even if the world steel demand and prices remain stable."[185] That was written in 2001 just as China's steel industry began its rapid expansion.

Surviving by Adding Value: Steel for the Automobile Industry

Surveying the future of the Canadian steel industry, Industry Canada offered this advisement: "Coupled with the continuing expansion of global steel production in developing countries and the high percentage of global steel production that is traded internationally, Canadian steelmakers are likely to face even more import competition in the future. This will be particularly true for low-priced, low value added commodity products where imports will first occur. Therefore, offering higher value added products and emphasizing service, delivery, and competitive prices as well as engineering and research and development capability will be important tools for domestic producers to keep one step ahead of the import competition."[186]

An ideal application of that strategy is the automobile industry. The world's automotive production in 2007 was 53 million cars and 20 million commercial vehicles.[187] The industry is concentrated among fifteen firms, which produce 92 percent of the world's automobile exports. The industry is also highly regionalized, with East Asia, North America, and the EU producing 80 percent of world output, and with the United States, Japan, and Germany producing about half. As wealthy consumer markets, the three regions also generate geographically concentrated demand.[188] The EU and NAFTA, where 75 percent of automobile trade is intra-bloc, encourage regionalization as well. These conditions also encourage economies of scale, to which automobile production is particularly well suited.[189] For an automobile assembly plant, scale economies are greatest at about 250,000 units per year, and for a steel mill producing automobile body panels scale economies can be as high as 2 million units per year.[190] Although the automobile industry shares the steel producers' exposure to economic cycles, as the 2008 global financial crisis showed when car sales declined enough to put the future of the Big Three in sudden peril, four characteristics make it an ideal customer: (1) its market size, (2) its need for long-term partnerships with nearby suppliers, (3) its continuing requirements for new high-technology steels, and (4) the opportunity it offers steelmakers to advance to component design and fabrication.

Big Customers

The automobile industry is the steel producers' largest buyer because of its size and the steel content of its products. The North American automobile industry consumes some 27 mmt per year, amounting to more than one-quarter of the steel producers' output, and a similar ratio prevails in Europe.[191] A challenge has been the automakers' concerns about fuel economy and vehicle weight, which have led them to substitute aluminum for steel. Between the late 1970s and 1990, the steel content of cars fell by 22 percent. Aluminum weighs half as much as steel, and some aluminum components, such as suspension arms, weigh only one-third as much. One Japanese supplier of aluminum suspension arms believes that efforts to reduce weight will increase the portion of aluminum components in cars from the current 6 to 10 percent to as much as 60 to 70 percent.[192] For the present, however, the trend has stabilized, with steel comprising 62 to 67 percent of average vehicle weight.[193]

Favouring steel over aluminum and plastics are lower costs – steel is about half the price of aluminum – ease of recycling, better surface finishes, and higher rates of production. Another advantage is the automobile industry's much longer familiarity with steel.[194] Cost, however, is the primary issue. According to one steel specialist, "automotive engineers have told me, 'I don't care what your material is as long as we meet the specs for the part. I won't pay a penny more than I have to ... Cost is king in the automotive industry now."[195] Overall, steel is "cost efficient when the combination of price per pound, strength, stiffness, and resistance to fatigue are considered."[196] Steel's biggest advantage is in the lower and medium automobile-product ranges, where car prices are the most competitive; the greatest substitution of aluminum is expected to be in the luxury car segment.[197] Steelmakers have been working to make their products more attractive. To counter the automakers' move to plastic fuel tanks, whose advantage is weight and formability, the steel industry has developed lighter steel units whose corrosion resistance is expected to extend to twenty years and that can meet new environmental vapour evaporation guidelines with thinner construction than plastic ones.[198] Even so, avoiding substitution will require steelmakers to meet evolving requirements at the lowest possible prices.

Long-Term Relationships

For components still made by the automobile manufacturers themselves and requiring only standard grades of steel, purchases can be made on the open market. One avenue is business-to-business Internet commerce. In going online to solicit steel bids, automakers have the advantage of size and buying power to shop for price and expect good results.[199] Steel that is more closely tailored to particular requirements, however, requires close, long-term, and interdependent ties between steel producers and automakers. For steelmakers coping with their characteristically volatile market, these ties offer stability and predictability. As the 2008 financial crisis showed, they also require a stable automobile industry.

Forging those ties has required a change in attitude. Historically, American steel producers, dealing with the automakers as commodity suppliers, had a "take it or leave it" outlook and were not responsive to complaints about poor quality and late deliveries. That changed in the 1970s and 1980s once foreign producers began entering the market. The steelmakers responded by demanding trade protection, but they also improved quality and dependability.[200] As they did so they came to appreciate closer relationships with the automakers. In economic theory this represented a major shift. According to the standard depiction, transactions take place in a market where prices, based on production and transportation costs, determine purchases. More recent formulations have added transaction and information costs. Lowering those costs and securing a dependable supply involves leaving the commodity market and entering long-term, "trust-based" relationships. When committed and specific resources are involved, those ties generate organizational interdependence as well as close and dense linkages. For the automakers the key values are dependable supply, consistent quality, and established channels for volumes of heavy material.[201]

Keeping transaction and information costs low means restricting participation. Ford and Chrysler are Stelco's largest customers, and Stelco is one of several suppliers to both. Smaller numbers are possible. In Japan, Nippon Steel and JFE provide 80 percent of Toyota's steel and 90 percent of Nissan's.[202] A tradeoff, however, is forgone alternative supply. In December 2004, Nissan Motors, after eliminating three steel providers in order to deal exclusively with Nippon and JFE, had to close three of its assembly plants for five days, even though the two steelmakers were running at full capacity. The result was delayed production for 25,000 cars. Nippon and JFE operate in world markets, and in December 2004 global steel supply was at its tightest owing to demand from China. To eliminate that market exposure, Hyundai Motors has decided to produce its own steel. Nonetheless, the pattern in the industry is for close relationships with long-term suppliers.

A practical consideration favouring those relationships is the automakers' large investments in processing machinery as well as tools and dies – investments that raise the cost of switching to different steel suppliers or different materials.[203] There is also a more qualitative consideration: the expertise and understandings that develop among closely linked firms. Related conditions favour proximity. Just-in-time inventory management, pioneered in Japan and adopted widely, has lowered costs because it involves holding only enough stock to keep operating. Although it is possible to run long supply chains, requirements for steady supplies and reliable transportation favour partners located nearby. So also, with steel, does sheer bulk volume. These have made proximity the norm in the North American, European, and Asian automobile industries. Tariffs are yet another consideration, and their absence under NAFTA is one factor in the two-way steel trade between Canada and the United States. Unpredictable swings in ocean shipping costs and foreign exchange rates also favour suppliers close to home.[204] Together, these are particularly immediate varieties of the advantages, seen in Chapter 4, that sustain industrial regions. In the closest automotive relationships, component suppliers connect their facilities directly to assembly plants.[205]

The nature of the automobile industry encourages interorganizational networks. In assembling cars, automobile manufacturers bring together output from three main production streams: engines and transmissions, bodies, and components. When the industry was vertically integrated that production was within the firm, but because these three streams are brought together only at final assembly they lend themselves to outsourcing. Along with other industries, the automobile manufacturers have reduced costs and simplified their operations by relying on specialized outside suppliers.[206] Leading the way in the decades following the Second World War were the European subsidiaries of Ford Motor Company and General Motors, which took advantage of geographic proximity to supply their British and German plants from neighbouring countries – a practice that developed in tandem with European economic integration.[207]

In Japan, the development of lean production, in which assemblers keep small supplies on hand and rely on just-in-time deliveries, required particularly close and continuous communication. The adoption of lean production by the other automobile manufacturers meant importing these interorganizational forms. The commodity flows have been accommodated by advances in logistics; the automobile industry is an important market for motor carriers, whose advantages over railways include flexibility and speed. The automobile industry is continuing to reduce vertical integration; today the trend is away from outsourcing components toward purchasing entire subassemblies. Today's automakers are evolving into system integrators; that is, more and more they are coordinating the activities of sets of suppliers. Long-term relationships encourage steelmakers to invest in specialized technologies and to tailor their products to the requirements of particular automobile manufacturers. That relationship, which stems naturally from close ties, was also pioneered in Japan.[208] Among Japanese automakers, collaboration with suppliers in materials and product development has saved time, lowered costs, and delivered better products.[209]

In recent years the North American and European automakers have been adopting Japanese practices. The interdependence involved in such relationships, and the experience and sunk costs that accumulate over time, are elements of long-term stability. The development of new steel products extends naturally into the design of new automobile bodies, opening the way to even more elaborate exchanges of expertise. As a consequence, the steelmakers are evolving into component producers, as will be seen shortly.

High-Tech Alliances
Specialized requirements are strongest at the top of the product line, where lighter and more crash- and corrosion-resistant vehicles demand innovative steel products. The strictest requirements are for external body panels, where flawless surfaces are of vital importance. The generic name for the new generation of automotive steels is advanced high-strength steel (AHSS). Characterized as a "diverse group of multiphase steels," AHHSs have double the yield and tensile strength of conventional automotive sheet.[210] The improvements are in formability, crush resistance, and energy absorption.[211] These characteristics meet the following design objectives: they make it possible to form complex shapes, they use less material without sacrificing panel and unit strength, and they absorb greater crash impacts.[212] The demands of weight reduction require that these qualities be provided by thinner steels. The change in steel content has been dramatic. Before the early 1990s, car bodies contained as much as 70 percent standard low-carbon steel. In contrast, 65 percent of the body structure of a Porsche Cayenne is AHSS.[213] The current Mercedes Benz C Class, introduced in 2007, has an AHSS content of 70 percent.

To pool development costs, expertise, and market access, steel producers in recent years have entered into alliances with one another. According to an OECD study, pooled R&D is one of the three most important reasons to enter alliances, the other two being joint manufacturing and production and joint sales and marketing.[214] Joint R&D provides the partners with access to one another's intangible assets, which include, centrally, technical expertise and experience in particular markets. This strategy is especially attractive when the costs of developing new products or production processes are high; when the market for them is dynamic, changeable, and competitive; when the customers purchase in high-volumes; and when they are globally dispersed.[215] Unlike formal mergers, alliances enable the partners to remain independent and to limit co-operation to particular activities. At the same time, because alliances are subject to less formal coordination, they carry higher risks and transaction costs.[216]

In Europe, Corus and Salzgitter Steel have developed high-strength and ductility steel (HSD). HSD steel combines great strength for crash resistance with high ductility for forming into complex shapes. It is "very strong and very elastic."[217] Another alliance, this one between Corus and Sumitomo, focuses on joint product development. Corus' interest here is supplying steel to Japanese auto plants in Britain; Sumitomo's is accessing Corus' technology and experience in dealing with the European automakers that are now expanding into Asia.[218]

A similar transregional venture is a product licensing agreement between Germany's ThyssenKrupp and Japan's JFE Steel involving an "exchange of recipes": ThyssenKrupp's CP (complex phase) steel for JFE's Nano steel. The agreement allows the two steelmakers to serve each other's automotive customers, thereby broadening the choices of suppliers, products, and buyers.[219] ThyssenKrupp and Arcelor have made a similar agreement to develop new "highly resistant and highly formable flat carbon" automotive steel (high manganese twinning-induced plasticity steel; TWIP). TWIP's advantage will be its 20 percent weight saving.[220]

Much of the steel research being conducted today involves collaborative efforts between steelmakers and automakers. In the United States, there is a jointly financed R&D partnership between the Big Three automakers and the steel producers. It was launched in 1988 to "study steel processing, develop new standards and design methods and make information freely available to suppliers." Its current focus is on designing parts and developing forming processes to allow greater use of thin-gauge high-strength steel. The three largest integrated producers have product development operations in the Detroit area.[221]

This statement from Arcelor's 2004 annual report highlights the multiple points of contact that are possible: "As the interface between the ... manufacturing units, research centers and automotive industry customers, Arcelor Auto is involved in every stage where steel is used, from the earliest design phases of vehicle projects through series production. The close relationship between its resident engineers and car design centers and its long-term business partnerships enable Arcelor Auto to offer steel solutions that successfully combine performance with safety, visual appeal, and economy."[222]

For steel producers, these contacts provide early and direct knowledge of the automakers' product plans and materials requirements – knowledge that can be fed directly into steel development. ThyssenKrupp's joint venture with JFE performs these functions with a marketing and consulting company that works closely with Japanese automobile manufactures in design and development. As an indication of the results that are possible through collaboration, ThyssenKrupp displayed the new Mercedes Benz R-Class sport utility vehicle at the 2005 Frankfurt Motor Show. Designed with Chrysler, the vehicle contains over 60 percent advanced high-strength steels.[223] Across the steel industry, working closely with automakers is becoming a "common practice."[224]

Another way to establish close ties with the automobile industry is through dedicated service centres. As was seen in Chapter 2, these installations – which are run either independently or by the steelmakers themselves – fill customer orders, fabricate steel in special forms and shapes, and apply special coatings and finishes. They are a significant presence in the industry. In the EU, some three thousand service centres, owned both by steel producers and by separate firms, process two-thirds of the region's steel, and automobile producers are among their clients. Eight of Arcelor's seventy-four centres specialize in serving the automobile industry. Their work has since expanded from managing steel supply and delivery into pre-processing for long-term partners. That work readily shades into manufacturing, as will be seen shortly.[225]

Smaller steel producers can stay abreast of current technology without joint ventures by developing close informal co-operation with automobile makers. When it was still an independent company, Dofasco mounted no special R&D projects, but it "strategically [adopted] best practices" in consultation with its customers, which include the automobile parts producer Magna International. External economies of scale facilitate these connections. As was seen in Chapter 4, external economies are generated by interconnected activities in industrial regions. Because of its location in the southwestern Ontario/Michigan automotive region, "Dofasco is surrounded by the type of tacit knowledge" needed for product development.[226] Under ArcelorMittal ownership, Dofasco will be part of a broader development network.

Japanese, European, and American automakers have production facilities beyond their home regions, and serving them has been a rationale for the alliances just seen. More ambitiously, a single producer can aspire to supply a steel product globally. Arcelor's 2004 annual report suggests the level of aspiration: "The powerful trends of economic concentration and globalization now being felt by many of the industries that consume flat carbon steel (especially the automotive and household appliance industries) are forcing major manufacturers to turn to suppliers that can supply them with identical products and services worldwide."[227]

Since then Arcelor has been acquired by Mittal Steel. Because the resulting company now has a presence in Europe, North and South America, and China, it enjoys the facilities and operational scope for individualized global supply. Acquiring a Japanese steelmaker would bring that manufacturing region under the ArcelorMittal aegis as well – a prospect, as was seen earlier, of which Japanese steelmakers are well aware. Such consolidation would expand the number of global suppliers.

Honourary Automobile Manufacturers

From supplying steel for the automakers to stamp into body assemblies, some steelmakers have moved into producing those assemblies themselves. This involvement has progressed in three stages. The initial stage involves cutting sheet into particular shapes, called blanks, to be stamped by the automaker into individual body components. In the second stage, steelmakers acquire stamping presses and perform the work themselves, producing tailored blanks – the formed steel parts. In the third phase, tailored blanks are welded into body modules. Once they begin performing the third phase, steelmakers become component manufacturers.[228] For the automakers, shifting that work to steelmakers saves capital and manufacturing costs and represents a shift of technology to steel producers. Across many component areas, the transfer represents a "de-integration" of automobile production.[229]

This development begets long-term relationships with steelmakers not only because of the specialized investment and experience involved, but also because the steelmakers possess knowledge of technological and product innovations that must be kept from competitors.[230] Close long-term relationships can also create dependence, however, and when there are few major customers and many potential suppliers the danger lies in becoming a price taker. That has been the position of auto-parts suppliers. When steel prices soared in 2004, they were unable to pass along their cost increases to the automakers, which were in a position to insist on low prices. Integrated steel producers are less dependent because they do have other customers, and the automakers are only one-quarter of their market. The American automobile industry's biggest steel supplier has been Mittal Steel, which has been a robust and resourceful counterpart, especially since its merger with Arcelor. Whether it will be *too* robust is a source of concern for the automotive industry, as was seen earlier. The relationship can be viewed as one between interactive oligopolies, whose dynamics were explored in Chapter 3.

European and Japanese steelmakers seem to be leading this transition. ThyssenKrupp has two centres for producing tailored blanks – in China with Anyang New Steel for blanks for Volkswagen's joint venture, and in Sweden at Volvo's plant at Olofstrom.[231] Another European steelmaker, Voestalpine, has moved into parts production; one of its products is completed door modules for the Porsche Cayenne. The firm's plan is to advance into automotive project management and engineering and to design and manufacture car bodies.[232] Among North America's integrated producers, only Dofasco reports that it is producing tailored blanks.[233] Blanking operations are performed by a number of steel service centres, and United States Steel, in a joint venture with Japan's Kobe Steel, produces coated steels for the automotive industry. The field at present looks to be open.

Fleeing Upmarket

The technology and expertise required to be in development and supply relationships with automakers are at the very top of the steel product line, and European and Japanese producers are at the forefront. Conceding lower-end products to imports and minimills has meant abandoning sizeable markets, but at the top end the automobile industry offers opportunities to add value not only by producing sophisticated steels but also by taking on activities the automakers once performed themselves. Since added value is reflected in prices, the high levels of added value in these processes are reflected in returns. This work is beyond the capacity of steelmakers in industrializing countries, and price discounting cannot substitute for missing content. A case in point is that the Japanese steelmakers have been using recent record profits for capital investment aimed at "cementing their positions as the world's leading producers of high-grade steel." This is largely in order to insulate themselves from the steel volumes being supplied by China and from softening prices in the world market for commodity steel.[234] Advantages in technology – as both physical and intellectual capital – provide that refuge.

Over the longer term, the 2008 global financial crisis may force the automobile industry to reorganize, but the production of cars will continue. If steelmakers can help the industry to simplify and lower costs, it is to the advantage of both.

Conclusion

The industry's sheer capacity makes its circumstances contingent. Forecasting errors and overzealous investment threaten to turn scarcities into surpluses and comfortable margins into losses. Unprecedented capacity, incentives to continue investing, unpredictable demand, and an unremitting burden of fixed costs together produce uneasy prospects. Although the world's demand for steel ensures that production will continue, the outlook for any particular producer is less certain, and none of the three survival strategies promises an escape from trouble.

What Kind of Future?

Protection is a variable and partial defence. As was seen in the Canadian and American practices, anti-dumping rules are administered by tribunals, whose decisions are based on investigations to establish the facts of dumping and injury. Although positive findings are indeed issued, the process is not predictable. And even when assessments are accurate, the two governments' willingness to protect steel may have changed over time. The benefits that do emerge are short-term, particularly in Canada's case with a five-year re-mission timetable, which underscores that short-term dumping is the kind warranting intervention. Demands for more all-encompassing policies arise from structural conditions that are beyond short-term remedies. Even when anti-dumping rules are more nominally regarded as an immediate measure – the official position of the steel industry – defence is neither automatic nor assured.

There is also the question of adequate remedy. The American steel industry did recover from the price trough of 2001, but as has been seen, it is rather doubtful that the Bush steel tariffs deserved the credit for it. One thing that certainly helped was capacity reduction, but that occurred because the industry drastically pared and recombined assets, many of which were in bankruptcy – hardly a tribute to full protection. Reorganization was largely complete by the time the tariffs were imposed. The Canadian government levied no equivalent tariffs, in part because it feared being overruled by the WTO.

A more positive option involves serving the automotive industry – an approach that plays to the established steelmakers' technological and or-ganizational advantages. The ongoing demand for strong and ductile steels is likely to continue, and as these materials become widely adopted in the automobile industry, the next phase of demand will be for low-cost produc-tion. At present, only the most advanced producers can meet that combina-tion – a comforting thought for established firms that are concerned about newer producers moving upmarket. An added benefit is that automakers are hospitable to arrangements that divide the tasks of production, use local suppliers, and require close coordination. Once established, these arrange-ments are too valuable to abandon casually, which promises the stability of mutual commitment. In addition, the automobile industry is expected to expand not only in China but also in India and Russia. Whether interdepend-ent relationships between steelmakers and automakers are transferable to those places remains to be seen.

Interdependence may still be uncertain since there are more steelmakers than automakers. The steel industry's inclination to consolidate does limit the number of candidates; that said, the automobile industry, through product-sharing alliances, shows the same inclination itself. Like the steel industry, it also has excess capacity. The automakers are big enough to wield powerful leverage on dependent suppliers, as parts and components manu-facturers can attest, and they may become more determined margin cutters if their conditions weaken. That prospect makes the next several years in-teresting to observe. Moreover, the fact that a successful relationship has been developed does not guarantee survival. Large sunk costs and valuable interorganizational ties may indeed accumulate, but suppliers can still be changed. Losing a vital account is a blow for any business, particularly if market conditions make the gap difficult to fill. For businesses carrying high fixed costs the situation readily becomes an emergency. Making such surprises more likely is the weakened health of some of the largest automakers and the prospect of drastic changes. Even if all of these concerns are resolvable in the steelmakers' favour, the automobile industry still constitutes only 25 percent of their sales. On the other hand, more than any other large purchaser, automakers demand what the successful integrated producers do best.

If the goal is stability without overreliance on strong relationships with buyers, consolidation has much in its favour. Agreements to cut output in the face of falling demand modulate the business cycle and limit price downturns. Pegging production to cover variable and fixed costs keeps finances stable until prices rise again. A related benefit of consolidation is that it exercises some force in one's own favour. Required, however, is a common interest in controlling volatility and a shared willingness to reduce production.

As was seen in Chapter 3, co-operation may be less straightforward in practice. First is the question of size. A membership large enough to affect prices is the minimal requirement, and stronger results require fuller participation. Each potential member must weigh the alternative of remaining independent. If enough do so, supply will be incompletely regulated. Being free to continue producing, in fact, is one motive for staying away. Also, the participants themselves may find co-operation difficult if they face sufficiently varied conditions in different markets. When price differences can be arbitraged between markets, members must wonder about the opportunities their co-operation creates. Complicating matters is the diversity of steel product lines, particularly when oversupply affects some products and not others.

In situations of worldwide downturns in steel demand, steelmakers may agree to production cutbacks in the early phase because their accumulated cash reserves can be expected to cover losses. That was the industry's stated expectation in the first months of the 2008 global financial crisis, and major producers did indeed cut output. They also said that they expected demand to recover in a matter of months. Willingness to continue cooperating may change if the crisis continues and cash reserves run out. That will be an important test of collective understandings.

Membership must also be small enough to coordinate. Unless the industry is reduced to around five members, verifying co-operation among numerous producers who operate facilities around the world and sell into dispersed and differentiated markets will always be a difficult and uncertain proposition. And even among a limited number, the classic prisoner's dilemma is readily imagined: the best individual outcome is enjoyed by one who keeps producing while all the others cut back.[235] The EU in the 1970s and 1980s faced that problem as it sought closings and rationalization from its members' steel producers. All knew there was too much obsolete capacity, and all knew as well that layoffs and adjustment costs would make governments reluctant to act, particularly if some faced bigger cutbacks than others. Quotas and adjustment payments were the solution, but they required single political authority.

There would be no equivalent of the European Commission to enforce compliance in a consolidated global steel industry, even though the individual incentives to avoid cutting back would be the same. As was seen in Chapter 3, cheating is a persistent problem in oligopolies. Ideally, all members would be guided by a collective benefit and would produce their allocated amounts, but at the same time, the prospect of extra sales would be a collective temptation. Weaker or less dependable members would require either some form of supervision or special incentives. Reliable data would be an issue, particularly if the industry were large and dispersed and if some members had doubtful incentives.

Maverick members would be a serious problem. OPEC's ranks include some independent-minded and financially needy governments, and accommodating them produces suboptimal prices. Even then, agreed portions are no assurance of full compliance. Unless the members' ability to withstand episodes of lower production and revenues is highly similar, the problem of relative losses arises and justifies the breaking of ranks by members with greater levels of suffering. Doing so shifts the cost of compliance to the other members. Beyond that, OPEC must bear additional reductions in order to compensate for large producers outside the ranks who choose either to keep on with current output or to expand production. These problems have been addressed in OPEC by the stronger members absorbing larger cutbacks. When they do so they provide the collective good of a coordinated price. A successful regime requires at least one major member that is willing to bear that responsibility. Whether it will do so dependably is a question that arises every time significant action is at hand.

Collective expedients all require fairly stable conditions. Given the amount of current and/or prospective excess capacity and its ability to exacerbate price swings, could the coordination available in a consolidated steel industry be more than a palliative influence? The question would be especially pertinent if a large non-member were to have divergent interests, or if a member government had its own unique incentives to produce more than an agreement stated. China comes most immediately to mind, but other countries might reveal themselves, given sufficient need or opportunity.

These considerations all raise the question of competition, which governments have the authority to regulate. In many countries efforts to coordinate production and manage prices are illegal. Canada's Competition Act, Part VI, states in Section 45.1:

Every member who conspires, combines, agrees or arranges with another person a) to limit unduly the facilities for transporting, producing, manufacturing, supplying, storing or dealing in any product, b) to prevent, lessen or limit, unduly, the manufacture or production of a product or to enhance unduly the price thereof c) to prevent or lessen, unduly, competition in the production, manufacture, purchase, barter, sale, storage, rental, transportation or supply of a product d) or otherwise restrain or injure competition unduly, is guilty of an indictable offence and liable to imprisonment for a term not exceeding five years or to a fine not exceeding ten million dollars or both.

The determinative standard is behaviour that "would or be likely to eliminate, completely or virtually, competition in the market to which it relates."

Article 81 of the European Community Treaty has the same provisions. Prohibited are agreements that "directly or indirectly fix purchases or selling prices or any other trading conditions" and that "limit or control production, markets, technical development, or investment." Article 82 applies the same standards to "any abuse of a dominant market position." In the United States, price fixing is a criminal offence under Article 1 of the Sherman Antitrust Act.

Enforcement hinges on what competition authorities treat as the threshold of permitted behaviour. In Canada, for example, activity must be sufficiently severe as to virtually or completely eliminate competition. In practical terms, establishing that level depends on the statutory interpretations of competition authorities. Investigating possible cartels, in fact, has been a priority of Canada's Competition Bureau, and coordinating steel prices may be more risky in this country than elsewhere. That makes the enforcement priorities in all of the jurisdictions where the steelmakers operate important and potentially changeable information. Significant differences in priorities would leave the strictest jurisdictions establishing the practical threshold. If that threshold were not high, then efforts among steel producers that were significant enough to counter a major drop in prices would also be significant enough to invite scrutiny and prosecution. Efforts below that threshold might be too little to have much benefit. If that were so, the co-operative potential in a consolidated industry might be easy to overstate.

- Competition law also allows governments to block mergers that threaten to restrict competition. On February 20, 2007, the US Department of Justice announced that it would require Mittal Steel, as a condition for the American government's approval of its 2006 acquisition of Arcelor, to divest its large integrated mill at Sparrows Point, Maryland. Previously owned by Bethlehem Steel, the seaport mill had passed into ownership of International Steel Group and then, when Mittal purchased ISG, of Mittal Steel. Tin-mill products were the market segment in question. A second option given to Mittal was to divest itself of Dofasco, which Arcelor had acquired and then placed in trust in an attempt to block the Mittal takeover. With that sale foreclosed, the Justice Department agreed that divesting Sparrows Point would resolve the issue.[236] In August 2007, the mill was purchased by a joint venture comprising Esmark, an American steel distributor that was also purchasing the steelmaker Wheeling-Pittsburgh Corporation; the Brazilian iron-ore producer Vale; and the Ukrainian steelmaker Industrial Union of Donbass Corporation. The new firm would import iron ore for Sparrows Point from Vale's Brazilian mines and send the mill's steel slab to Wheeling-Pittsburgh's mill in Ohio for processing. The Ukrainian steelmaker would provide an additional source of slab.[237] In 2008, the mill was purchased by Severstal. Given the size of the assets and corporations in question, this episode highlights the legal power available to governments to maintain competition.

In light of the limitations of all three strategies – protectionism, consolidation, and close alignment with the automobile industry – survival is still a hedged and contingent proposition. It well merits characterizing steel's integrated sector as expensive, volatile, and difficult. A mitigating influence, surprisingly, is globalization, as will be seen in the concluding chapter.

6
Steel in a Global Perspective

We have seen a product that is a mainstay of modern economies, that is made in standardized grades and is widely traded. It comes from an industry that is vulnerable to oversupply and weak prices because each member's incentive is to increase production and revenue. If others do the same and create a surplus, or if demand falls in a cyclical downturn, each one's incentive is to cut prices. When that happens all are worse off, and the weakest fail. These conditions are rooted in the industry's technology and cost structure, and they imbue it with volatility and risk.

When surpluses occur in one country and shortages in another, price differences encourage trade. They also allow excess supply to be unloaded in others' markets. Discounts can do the same, enabling foreign producers to gain market share. Because trade allows these forces to operate in multiple directions, and because technology and cost structures are an equalizing influence in the industry, producers in all locales desire market stability. That which is available from trade protection, from oligopolistic co-operation among consolidated producers, and from bonding with primary customers is incomplete, leaving producers on their own in open and contestable markets. At the same time, trade enables steel's consumers to enjoy diverse supplies and competitive prices. Trade over long distances is possible because of economies of scale and efficient transport. Current steelmaking technology provides large economies of scale, and transportation costs have made transoceanic supply feasible. These are the elements of a globalized steel industry, and what that portends is the subject of this chapter.

Blended Entities

A common notion of globalization is that it makes things homogenous, and that characterization amply applies to steel. Commodity-grade steel is produced to common international standards, making it is an excellent example of a homogenous good – a class of articles that buyers see as fully interchangeable. That has long been true among established steelmakers and reflects the evolution of industrial practice in Europe and North America. New producers have adopted the same standards partly because they are embedded in steelmaking technology and partly because steel's most common products incorporate them.

To the extent that successive rounds of trade liberalization have increased access for imports, the international steel market is also homogenous. Standardization and substitutablility eliminate qualitative barriers to entry into foreign markets and, for commodity grades especially, make price the point of differentiation. That fact has strong implications for producers, as was seen in detail in previous chapters. What is important here is the degree of homogeneity that makes prices so determinate. The same standardization establishes parameters for the design and content of manufactured goods; and producing them for international markets, particularly as part of integrated supply networks, requires dependably uniform materials. More generally, the current surge in steel demand has come just as industrializing economies are transforming themselves along the lines of the established ones – in infrastructures, durable-goods markets, and manufacturers' requirements.

Steelmaking technology is standardized, portable, and commercially available. That has enabled countries with no previous experience to establish themselves as steel producers; it has also enabled the industry to spread into new regions. The same portability has enabled countries whose steelmakers had served local markets and used old technology to modernize rapidly. In both cases, buyers of the technology can begin exporting. Standardization yields products that can be sold immediately throughout the world market, saving the time and investment required to start afresh. And because technology contains embedded knowledge, that knowledge also spreads and becomes homogenous. Engineers and metallurgists apply that knowledge and develop it further.

Finally, the industry itself is homogenous in the way it functions. Producers use the same technology and organize production in the same kinds of segmented installations. Production processes are shared throughout the industry – a tribute both to technological standardization and to evolved, reproduced, and professionalized practices of engineering. The same basic routines, organizational structures, and automated processes manage production. On the input side, steelmakers that do not own their iron-ore sources buy from the same mining oligopoly at world prices.

These features have characterized the industry in its historical and current states. They have evolved with technology, markets, and product standards and are not particularly new. What *is* new is global consolidation, which is placing formerly independent producers under larger corporate structures of ownership, direction, and control. New as well is the entry into world markets of producers in industrializing countries.

Globalization and the Costs of Trade

With these impressively homogenous elements already in place, what further enlargements and extensions could there be? There are two possibilities, both of which proceed from the costs of international trade. In the first, those costs stay the same. With no further advantages to be gained from selling steel internationally, the integrative possibilities in steel's shared technology, product standards, and markets remain at their present levels, although ownership may continue to coalesce. From the perspective of the changes that produce globalization, the industry has reached a plateau, and stable costs of trade mean that it is likely to remain there. In the second possibility, the costs of international trade fall to the level of the costs of inter-regional domestic trade. That could encourage a radical reorganization of the industry, with facilities that now serve national markets specializing fully in order to serve global ones. This would particularly favour those large producers that possess (a) the required scope of ownership and coordination to organize globally and (b) the resources to invest in fully specialized facilities. That change would remove the last vestiges of the era when steel producers and markets were primarily national. From a global perspective, it would represent the ultimate disappearance of the boundaries that separate producers from markets. The incentive for reorganizing is to reap at a fully global level the advantages of specialization: economies of scale, lower costs and prices, and vast potential markets. The result would be a homogenous worldwide expanse.

Under either possibility, the existing division between producers in industrializing countries specializing in commodity-grade steel and established producers specializing in specialty-grade steels would simply extend globally. If trade costs were to remain unchanged, both sets of producers would divide their efforts between home and export markets, and production and trade patterns would remain much the same as now. If trade costs were to fall, both commodity and specialty steelmakers would be able to select their most competitive items and devote their resources to them. Unless the world industry consolidated and specialized to the point where a handful of global producers became exclusive suppliers of particular items, competition would remain. Both commodity and specialty producers would sell in home and overseas markets and expect other producers to do the same. For commodity-steel producers, high interchangeability would continue to make prices determinate, creating an incentive to achieve even greater scale economies and favouring steelmakers with already large outputs. For producers of specialty grades, differentiated products would create incentives to fund research and development, invest in advanced production facilities, and cultivate connections with major buyers.

These two possibilities were proposed in an econometric study of world trade that tested two comprehensive models – comparative advantage and economic geography. Under the comparative advantage model, trade arises from complementary factor endowments. Countries that enjoy endowments that make them efficient in one good trade with countries that enjoy endowments that make them efficient in another. The respective cost differences generate trade through arbitrage – that is, goods that are cheap in one locale are transported and sold in the other locale where they are expensive. Under the economic geography model, trade arises from economies of scale and low transportation costs. There, too, cost differences generate trade. Economies of scale make goods cheap to produce, and low transportation costs make it feasible to retain that benefit in distant markets. In both models, successful trade encourages further specialization.

Among the OECD members studied, the results showed that, although some industries did show the presence of scale economies and transportation costs as encouragements to trade, overall the "economic geography effects were not robust."[1] However, applying the same analysis to inter-regional trade data from Japan showed that in eight of nineteen manufacturing sectors, including iron and steel, there were statistically and economically significant economic geography effects – that is, economies of scale and transportation costs did indeed account for trade flows. The authors' tests of their variables persuaded them that they had properly identified and measured the economic-geography factors and that the international and inter-regional comparison was valid. In addressing the question, "Why are the regional effects of economic geography so strong, while the international effects are so weak?" the authors proposed two explanations. The first is that trade costs, which include transportation but also all the other impediments to the movement of goods, "must surely be lower for trade between regions of a country than between countries." The second is that factor mobility is greater within than between countries. That makes it easier to concentrate production where scale economies are the strongest and, with favourable transportation costs, to supply other regions. The implication is that, at the international level, trade costs are still very strong – enough to "overwhelm" economic geography factors.[2]

Economist John Helliwell has devoted much study to the question of why trade flows much more heavily within countries than between them. One purpose has been to identify border effects – the conditions that discourage international commerce. These conditions subsume costs of trade, but Helliwell considers them more broadly. Their influence is striking. For industrial countries border effects are "significant" and for developing countries greater still. They are strong even among long-time members of the EU, where efforts to eliminate trade barriers have been comprehensive. Among EU members not possessing the facilitating condition of a common language, their domestic trade is *six times* more dense than their bilateral trade (emphasis mine). The pattern, moreover, is not sectoral but generalized across industries. Outside the EU, the usual suspects of non-tariff barriers and other restrictions are not the reason why the level of international trade is so much lower than domestic trade. Policy disparities among trading partners are not a major impediment either.[3] Instead, in Helliwell's view, border effects stem from the costs of information and transportation, which expand with distance. More important still are qualitative differences – among laws, commercial and informational networks, and local practices. These differences expand with distance and even more with dissimilarity. The least costly trade is within established and dense commercial and social networks, and these proliferate most readily and fully within national jurisdictions.[4]

How do these conditions apply to trade in steel? At first glance, "soft" factors such as culture and commonality would not seem to fit a "hard" product such as steel. Robustly utilitarian as both an industry and a commodity, steel would seem to be unaffected by qualitative and contextual differences. As has also been seen throughout, most steel is readily substitutable. To foreign buyers, that eliminates uncertainties about quality that would otherwise need to be resolved in familiar commercial networks, where local suppliers would indeed have an advantage. Substitutability also provides insurance against unreliable delivery, since alternative supplies are often readily available. These considerations make dealing with a distant or unfamiliar vendor less risky. Altogether, steel's specific and objective qualities and applications would seem to limit the need for familiarity and close connections.

At the same time, there are conditions in the steel trade that do favour producers that are known through networks and practical experience. Primary customers such as automobile manufacturers require very dependable quality and delivery. Operating on just-in-time logistics makes them particularly unable to tolerate late deliveries or bad goods, since no reserve stocks are kept on hand. Although automakers are able to meet some of their needs on the open market, for steels that must fit very specific requirements they prefer working closely with local mills. Also relying on dependable vendors are construction firms, which work on critical-path schedules with precise and sequential delivery times. They, too, are unable to keep large stockpiles because of limited capital and space and the individualized steel requirements of particular projects. Construction firms and many manufacturers get their steel from service centres, which in turn buy their supplies from the mills. That fact simply takes the requirements back one step: as distributors, service centres rely on known and dependable suppliers. The same requirements – especially punctuality – impinge on transportation firms, which figure directly in the economic geography explanation of trade and which rely considerably on reputation in soliciting time-sensitive goods. The trust that underpins these relationships develops through contacts, familiarity, and proximity. In this light, steel producers that have established themselves in foreign markets and that have gained experience in dealing with different laws and practices possess valuable intangible assets. These kinds of network factors — especially proximity and familiarity — do much to explain the large and active two-way steel trade between the United States and Canada. Moreover, the countries' steelmakers have operations on both sides of the border, and these further enhance network advantages.

When these considerations are highly important, they may readily override cheaper prices. That is one reason why exporters, especially ones seeking to penetrate new markets, rely on discounts. When experience and trust matter, discounts can be seen as a bribe to buyers who would otherwise prefer dealing with familiar counterparts. These considerations also help explain why buying foreign steelmakers has become popular: for producers seeking new foreign markets, these acquisitions provide not only the hard assets of physical facilities but also the soft assets of established positions in local networks.

These conditions need not be static. Producers that succeed in new foreign markets do so by providing good products and service. Over time they build business reputations and become known in commercial circles. For producers long established in Europe, North America, and Japan, reputations may provide advance credibility elsewhere and be quite portable. For outsiders, that cachet can be acquired by merging with a major producer. That was an important motive in Mittal's purchase of Arcelor, Tata's purchase of Corus, and Essar's purchase of Algoma. In the same way, reputations in foreign markets can also be bought, with Arcelor's established presence in Brazil accruing to Mittal. The same prospect attracts buyers from Russia and, potentially, ones from China.

A Stable Hierarchy?

Steelmakers in Europe, North America, and Japan have conceded much of their commodity-steel business to imports and minimills. That has created a stratified order, with steelmakers in industrializing countries exporting finished commodity products such as steel coil and plate as well as semi-finished ones such as slab. Established producers in Europe, North America, and Japan devote their efforts to the top of the product line, where their research capabilities and specialized technology can be put to optimal use. Is this division of labour likely to continue? Movement into the ranks of specialty producers would require investment in technology, expertise, and reputation. At the top of the line, where sophisticated and expensive human and technological capital is required to work with complex metallurgy, formulate multiproperty sheet steels for demanding customers, and participate in new product design, barriers to entry are significant. Would any commodity-steel producers wish to pay those expensive admission fees?

There are several reasons why they would. First, the world's automobile industry, the primary customer for fine specialty steels, is already large, and in markets such as China and India it is expected to grow rapidly. As markets in those countries become wealthier and more discerning, they will begin demanding vehicles containing higher-quality steels, and locally designed vehicles will begin incorporating them. Automakers producing vehicles designed in Europe, Japan, or North America will simply follow those specifications in their new markets. These trends are creating opportunities for all steelmakers, but particularly for those already established locally as commodity-steel producers, for whom an expanding automotive sector provides an incentive to add sheet-steel facilities. Second, the markets for the broader array of specialty steels are now mainly in the richest countries but prospectively in those where wealth is rapidly propagating. One reason is that economic advancement supports the kinds of manufacturing – of medical equipment, for example – that use specialty steels. Third, to the extent that status in an industry attaches to making products high in added value and technical content, there is an incentive to join in. The results may not equal those of the industry's leaders, but the effort places newcomers in the leading sector – provided that in seeking that status they do not lose profitability. Finally, there is the more ambitious objective of actually displacing respected incumbents. Vulnerable ones would be those that are unable to afford product development and new technology – or that simply become complacent – and begin to lag behind the leaders. The difficulties of troubled steelmakers staying abreast of technology were seen in Chapter 2. Whatever combination of inducements actually initiates upward mobility, successful moves would rearrange the industry.

How will ambitious steelmakers ascend the ranks? The basic means can be seen in Kaldorian strategies (named after the economist Nicholas Kaldor). As a means of advancing in an industry, these strategies are particularly relevant for newcomers lacking experience, or the assets of comparative advantage, or both. The strategy involves acquiring efficiency and competitive position by vigorously using productive assets, either existing or newly acquired. That involves seeking economies of scale, acquiring proficiency through learning by doing, and perfecting production processes. These results require output to be increased. Doing so captures scale economies, and the greater volumes accelerate the mastery of technology and production. As that happens, productivity rises, generating the resources for investment in better technology and new products. The strategy is based on cumulative benefits, in which initial gains set the basis for subsequent and greater ones. The objective is to become the most proficient producer, reap the benefits of income and market share, and use the proceeds to advance further. If those efforts succeed, the firm will be in a position to challenge incumbents with lower prices and better products.[5]

The risk is that these transformations will not occur and that investment will be wasted. That can happen because the key processes are not so much technological as social: they "emerge from the interaction of managers and workers on the factory floor. Management may be incompetent and fail to recognize ways to increase throughput or cut costs, and workers may be powerful enough to resist reorganization of work processes." For a firm the consequence is cumulating losses. For a state sponsor, "subsidizing exports may just run up a big debt, while subsidized firms fail to learn."[6] Firms that succeed, however, may build up the momentum to enter new markets, buy new technologies, and develop technologies of their own. The risk that one's efforts will fail may be avoided simply by buying an already successful firm. That activity has been seen throughout this discussion. What is useful to note here is that it represents a different use of Kaldorian proceeds.

Buying a more advanced firm and appropriating its technology can be seen as an internal transfer. Technology may also be acquired externally, and there are three ways of doing so: direct purchase, transfer from a parent company, and joint ventures. It is easy to assume that simply buying and installing new technology will bring improvements, but according to a recent econometric analysis of technology transfer in Chinese industries, that alone is not enough: directly purchased technology "exhibits no effect on firm performance." To benefit, firms must have their own R&D capability. In the authors' words, "R and D and technology transfer are not substitutes; they are highly complementary ... when combined with R and D, foreign technology transfer generates measurable productivity gains." The benefits do not end there. Adopting new technology also increases returns to the firm's R&D. That result "underscores the critical role of in-house R and D capabilities as an important channel for absorbing externally acquired technologies."[7] It also underscores that one requirement of Kaldorian strategies is an autonomous capacity for generating new processes and goods. China requires foreign joint-venture partners to transfer technology as a condition of setting up production in the country. That further accelerates a firm's advance because the transferring partner loses exclusive control over its technology and opens market space for its venture partner. That prospect constitutes a major disincentive for setting up production in China.[8] When the receiving partner already has R&D capability the disincentive is even greater.

This raises a question: Which industrializing-country steelmaker is best positioned to enter the senior ranks? For the reasons just seen, that firm is China's Baosteel. The Kaldorian requirements are all present: China's huge and voracious domestic steel market can amply generate the volumes needed for scale economies. And as China's manufacturing sector makes the transition toward producing goods with higher technological and value-added content, it will become a growing market for high-end steel products, supplying the necessary volume there as well. The transition accords well with Chinese government policy, which was first to reduce imports of commodity steel and then those of specialty steel. Although most of Baosteel's products are consumed domestically, it has begun exporting surpluses and, as was seen earlier, contemplating opportunities in North America. Its rate of growth – 27 percent in 2006-7 – continues to provide ample scope for economies of scale.[9]

From the beginning, Baosteel has been developing its own R&D. Although it purchased its initial plant from South Korea and Japan, these were joint ventures, and equipment for the company's new hot- and cold-rolling mills was "co-manufactured" with foreign partners. More impressively, Baosteel used home-built technology for its "blast furnace, coking furnace, sintering machines and hot and cold strip mills." Altogether, Chinese-made equipment amounted to 88 percent of the project. That was in the early 1990s, and even then, Baosteel's technology was "among the best in the world," showing "the potential of the Chinese steel industry to upgrade its technology via technology transfer."[10] Much more recently, as was seen in Chapter 4, Baosteel has entered into joint ventures with Arcelor and Nippon to produce automotive sheet steel. That project has transferred advanced metallurgy and production technologies. The joint venture was prompted by the rapid growth expected for China's automotive sector. Given Baosteel's existing level of R&D, the complementarities seem to be impressive.

A Global Blend

According to a standard view of economic development patterns, producers in countries such as China, Brazil, and India fit the profile of late industrializers. Those operating in limited home markets depend on exports to achieve the necessary scale economies to advance. That pattern has been somewhat true in steel, although all of the newcomers, as was seen in Chapter 4, were originally intended to supply their countries' own domestic industries. Competition from imports and domestic minimills has led established steelmakers to concede markets in lower-level commodity product lines. Those concessions can be a net benefit to the home economy because they open the way for cheaper goods. Wealthy and mature economies adjust by abandoning those industries according to the logic of comparative advantage; that in turn allows industries with higher labour content or low levels of value added to migrate to new producers. From a welfare standpoint, the ideal outcome is using the savings from cheaper goods to allow conversion to products that require more expensive and elaborate capital and that contain higher added value. That transition is especially possible in times of economic growth and prosperity because new leading sectors are more readily financed and labour force adjustments are less wrenching.[11] To the extent that established steelmakers in Europe, Japan, and North America have conceded product lines, shed inefficient capacity, and concentrated on the upper levels of the market, that adjustment fits the standard view.

What is different is bottom-up mergers in which steelmakers in industrial-izing countries buy established firms. Tata Steel's acquisition of Corus was greeted with public jubilation in India: it represented a massive entry into a large and wealthy market; it involved the purchase of advanced technol-ogy; and, in an ironic post-colonial twist, it meant that in effect, India was acquiring the assets of British Steel. Mittal, when it purchased Arcelor, and Tata, when it purchased Corus, were emphatic in their assurances that these were valued assets and would not be stripped. The purpose was to own and operate them. There are good reasons for keeping things in place. One is the very large sunk cost in expensive and sophisticated facilities. Another is their location in the largest and wealthiest markets, with all the advantages of established network positions. Yet another is that all are self-sustaining and do not require cross-subsidies from their owners. Acquisitions can go beyond simply purchasing mills. When India's Essar acquired Algoma in 2007, it also purchased Minnesota Steel, which owns iron-ore–producing facilities in the Mesabi Range. Besides financing a new steel mill on the site and processing ore to supply Algoma, Essar made a purchase offer for Esmark. Based in Wheeling, West Virginia, Esmark operates eleven steel-service centres in North America and is rapidly expanding. "With Esmark's distribution network, Essar can now take the Algoma products deeper into the market. Further, Esmark produces a range of products, including galvanizing [in original] steel for high-end consumers. Thus, Algoma steel can be shipped to the Esmark facility for galvanizing and pushed to the market through the latter's distribution channel."[12] The Esmark purchase, as was seen in Chapter 4, did not succeed, and the company went to the Russian steelmaker Severstal, which wanted it for the same reasons. Money to acquire European and North American steelmakers began to flow in 2004 – a banner year for steel prices – and continuing high demand on the world market would keep producers in a position to buy.

If this pattern continues, the industry's present hierarchical differentiations will meld into a global blend. That outcome will differ strikingly from depic-tions of globalization in which wealthy firms in the North appropriate fledgling ones in the South; in which wealthy states extract resources and profits from developing ones; and in which producers in the South are rel-egated to labour-intensive and low value-added tasks. A blended steel industry would also foster an unexpected form of interdependence. The health of these firms would depend on conditions in a set of dissimilar markets. Being present in all can be seen either as a globalization of shared benefits in good years and of shared perils in bad ones, or as distributing risk over a set of separate venues. Over time, as markets evolve and steel firms rearrange their technology and facilities, an amalgam could emerge that entails fused owner-ship among wealthy and industrializing countries, with goods produced according to local demand and export opportunities, and with full arrays within individual firms of commodity-grade and specialty steels.

How do these considerations appear in light of the costs of trade? If those costs do not fall, the industry's level of international involvement will remain at its present plateau, although the firms will be more cosmopolitan. Producers will be acquired and organized according to their position in existing national markets, and exporting will continue according to present patterns. With all producers doing the same, it will be difficult for any to dominate entire global market segments. Export opportunities will continue to depend on price levels and on local surpluses or shortages. That will further limit incentives to specialize, even where large structures of ownership and control make it possible.

On the other hand, blended firms may lower some costs of trade. Since these firms would be large enough to be present in multiple markets, their size and global reputations would remove many of the qualitative barriers to trade just seen and facilitate more widespread commerce. If it is true that border effects are as much informally qualitative and associational as they are formally governmental and regulatory, they would yield to firms composed of once-familiar local producers. As firms incorporate their new members' network connections and generalize them across markets, reputations themselves may become homogenous. Limiting the global scope of unification would be the remaining independent advanced producers, such as ThyssenKrupp and JFE, merging defensively with one another and not including steelmakers from outside the circle of advanced economies. Even then, with other large and more diverse amalgams present, much of the blending effect would accrue.

The Economic Geography Model Undone?
The economic geography model explains trade as a combination of economies of scale and transportation costs. Scale economies lower unit costs, make products price-competitive, and encourage specialization. Transportation costs enable efficient and specialized producers to serve distant markets. Chapter 4 showed how these conditions encouraged the dispersion of steel production beyond its historic centres in Europe and North America. As was just seen, the advantages of specialization and favourable transportation costs may not be enough to overcome barriers to trade, including qualitative ones based on commercial networks and familiarity.

2008's soaring energy costs raised a more fundamental prospect: that transportation will become increasingly expensive, particularly over great oceanic distances, offsetting export advantages and favouring local and regional producers. For steel, these effects are enhanced by weight and relatively low unit value, which together make steel expensive to ship.

Transportation in all modes is strongly and immediately affected by fuel costs. These translate directly into operating costs, which must be covered by shipping rates. In a widely cited study, CIBC World Markets' chief economist Jeffrey Rubin tracked oil prices and transportation costs. At 2001 oil prices, averaged over all transportation modes, fuel costs represented 20 percent of operating costs. At $150 per barrel, fuel costs rise to 50 percent, and at $200 they rise to 80 percent. Since 2005, each dollar increase in oil prices has "fed directly into a one percent rise in transport costs."[13] The same is true with distance: At 2008 oil prices, "every 10 percent increase in trip distance translates into a 4.5 percent increase in transport costs." To show the direct implications for trade, Rubin converted transportation costs into their equivalent rates of tariff, using those of the United States as a benchmark. Even at the $20 per barrel price prevailing in 2000, transportation costs equalled a tariff rate of 3 percent. Transportation costs in the first half of 2008 raised the equivalent tariff to 9 percent. Thus, $150 oil would equal an 11 percent tariff. "At $200 per barrel," notes Rubin, "we are back at 'tariff' rates not seen since prior to the Kennedy Round GATT negotiations of the mid-1960s."[14]

Chapter 2 showed the effect of rising shipping rates on the price of iron ore delivered to China. Because of the distance involved, sea transportation accounts for 50 percent of the cost of Brazilian ore landed in China, while Australian ore, with a shorter journey, includes only a 30 percent transportation cost. That difference, plus tight supply and China's growing demand for ore, enabled Rio Tinto to extract from Baosteel a hefty premium over the marker-price increase of 65 percent that Vale had just concluded with Nippon Steel and POSCO. Pricing power notwithstanding, however, transportation costs do rise with oil prices and distance. These translate directly into costs of production.

The same constraints are present on the export side. At an oil price of $20 per barrel, the cost of shipping a standard container from Shanghai to the east coast of the United States was $3,000. At 2008 oil prices, that rose to $8,000, and at $200 per barrel, the cost would rise to $15,000.[15] Historically, Rubin showed, trade volumes from Europe and Asia to North America fell as direct results of the oil shocks in 1973 and 1980, even though economic recoveries following the shocks should have restored previous levels. Instead, trade lagged the recovery because the shocks had tripled transportation costs.[16] Prospects for steel exporters supplying distant markets are not promising. At summer 2008 oil prices, a tonne of hot-rolled steel costs $90 to ship from China to the United States.[17] Transportation costs such as these must be offset by lower production costs or discounted prices to make goods competitive in export markets. As was just seen, however, China also bears high shipping costs for its imported iron ore and coal and deals with formidably powerful suppliers. North American steelmakers are free of both constraints. In addition, China's labour cost advantage, as was seen in Chapter 4, has been shrunk by automation. Lower trade volume reflects these facts. Rubin reported that China's steel exports to the United States were falling at an annual rate of 20 percent, while American producers were expanding production at an annual rate of 10 percent. Across all the categories of Chinese goods that are expensive to ship, the transportation-cost penalty represented a 30 percent reduction of expected growth in the American market.[18] These implications do not eliminate the prospect of global steel gluts and bouts of destructive price cutting. They do mean that offsetting higher transportation costs would require deeper discounts.

From a global perspective, high oil prices will divert trade from remote producers to more proximate ones. For North America, Mexico stands to supplant lost volume from China, and the change will be the most striking with goods bearing high shipping costs. Instead of travelling on the sea they will move overland, and a prime beneficiary will be the Kansas City Southern Railway, whose routes extend from the American Midwest into northeastern Mexico. A major change expected would occur in the automobile industry, with production being diverted from China.[19] Prospects differ for steel because only 26 percent of Mexico's steel production is in the integrated sector, compared to 41 percent for the United States and 59 percent for Canada.[20] One implication would be an even greater bonus for American and Canadian steelmakers. The same diversion occurred in other industries. Tesla Motors, which is producing a battery-powered car to sell in the United States, had originally planned on a global supply chain, with the battery packs being made in Thailand, then shipped to Britain for installation in semi-finished vehicles, which would then be shipped the United States for final assembly. Instead, when Tesla began production in 2008, it decided to switch both operations to its location in California. Doing so would save shipping costs of 5,000 miles per car. According to Tesla's senior vice-president of global sales, marketing, and service, "it was kind of a no-brain decision for us. A major reason was to avoid the transportation costs, which are terrible."[21] The same incentive would shift Tesla's steel supply. In another move, the Swedish furniture maker IKEA, which epitomizes global scale and expanse, opened its first American factory in 2008. The implications bear more generally on supply linkages and industrial clusters, with one result being a "neighbourhood effect" – that is, interdependent producers relocating closer to one another.[22] Since one of the economic-geography model's two pillars is favourable transportation costs, these effects undermine its expansive implications. "Globalization," in Rubin's words, "is reversible." Following their peak in the late spring of 2008, world oil prices dropped by two-thirds by December in response to the prospect of a prolonged global recession. If oil prices remain low, the transportation pillar of the economic geography model will remain in place. If they rise again, the changes just seen could recur. If they then remain high, the changes could be long-term.

The State Redux

An unexpected prospect is the return of the state – this time foreign states. As was seen in Chapter 1, governments in Europe and North America have begun to worry that trusts affiliated with foreign governments will make large investments in their economies. Russia and other oil-exporting states in the Middle East, along with countries such as China and Singapore that hold large trade surpluses, have amassed vast stores of money. At present these funds are estimated to hold some $2.5 trillion in assets, and at current rates of accumulation they could grow to $17.5 trillion within a decade.[23] When they acquire large blocks of portfolio investments or purchase firms outright, they raise two concerns. The first is that, even acting solely as asset managers, they may have a strong impact on financial markets. They could cause turmoil by dumping assets or, alternatively, add stability in times of crisis. The second is that they may use their position to pursue their interests politically by demanding regulatory changes, for example, or other special benefits. Those efforts would go beyond normal lobbying because the background presence of a foreign government would raise the question of interests beyond the purely commercial. Even if the foreign government's influence were not actually exercised, it would still be implicit. Exercising explicit influence, governments could manipulate their assets for political ends by coupling demands with the prospect of disruption.

More imposing is the question of strategic sectors. As was seen in Chapter 1, states historically had treated steel as a strategic industry because of its importance in manufacturing and particularly in defence production. Countries of the EU have been concerned that Russian energy companies buying European gas distribution networks would enable the Russian government to threaten supply cut-offs. Aware of that possibility, the Czech government in the early 1990s constructed a pipeline to Germany as an alternative to the Druzhba pipeline from Russia. The move paid off in July 2008 when the Russian pipeline monopoly Transneft disrupted oil supply to the Czech Republic immediately after the government signed an agreement for joint missile defence with the United States.[24] These considerations raise the question of what now constitutes a strategic industry and how acquisitions affect national security. For industries that develop and control highly sensitive technologies, the question is straightforward, but for others exposure is less obvious. There is also the task of distinguishing between clearly commercial state-backed investment funds, such as Norway's, and ones with less transparent purposes and operations, such as China's.[25] The Russian bank Vneshtorgbank caused unease in Europe when it acquired a 5.4 percent stake in EADS, the aerospace firm that controls Airbus Industrie. Although the bank denied that the move was anything more than commercial, the hand of the state appeared when senior officials and President Vladimir Putin stated their desire for a Russian role in Airbus.[26] Even when there is no effort to disguise involvement, such acquisitions still mean using state resources and power to establish positions in key sectors.[27]

Political ties exist in the Russian steel industry. In fact, some Western investors interested in Russian steelmakers see close relations with the Kremlin as an advantage because of reduced political risk.[28] The degree of risk for firms not enjoying political favour became apparent in 2008 when Prime Minister Vladimir Putin, as was seen in Chapter 4, accused the Russian metallurgical company Mechel of "incorrect price policies" and directed the government to investigate why the firm was exporting metals at a price lower than the one charged in Russia. Share prices of Mechel fell sharply in response. The business community in Russia saw this as evidence that firms are not free from the Kremlin's interference and manipulation.[29] The purchase of Western steelmakers by state-backed investment trusts or politically affiliated corporations is not a fanciful prospect. OAO Severstal made an unsuccessful bid for Stelco in 2005 and in the same year entered a venture to establish SeverCorr, a new American minimill serving the automobile industry, acquiring full ownership in 2008. The Russian steelmaker Evraz purchased the minimill Oregon Steel in 2007 and announced its interest in buying the minimill Ipsco. The impeding factor turned out to be possible antitrust action by American regulators, since Oregon Steel and Ipsco would "dominate the markets for plate steel and large-diameter pipe." The "hammer" in this case, then, was competition law.[30] Ipsco, which has facilities in Canada and the United States, was purchased by the Swedish steelmaker Svenskt Stal AB in July 2007, and in June 2008 all but one of Ipsco's five Canadian plants were sold to Evraz. Since one of Ipsco's main businesses is serving North America's petroleum industry, the deal involves a key sector.

A question is whether such deals also involve national security. Evraz's purchase of Oregon Steel was reviewed by the US government because one of the firm's products is armour for the military.[31] Although American law allows the barring of foreign takeovers that endanger national security, the sale was approved with little delay. Subsequent legislation, the Foreign Investment and National Security Act, has extended review to government-backed investment trusts. The German government, also attentive to security, has used the American legislation as a model for its own.[32] In Canada, the Competition Policy Review Panel was formed in June 2007 to examine whether the Competition Act and the Investment Canada Act need to be revised to cover state-owned entities and national security. In October 2007, the government removed national security and state-owned enterprise considerations from the panel's mandate; the report, issued in June 2008, focused instead on liberalizing foreign investment rules.[33] Left open is the possibility that the government will address strategic questions with separate legislation. Concern about ownership in the steel industry may no longer be a relic of an earlier geopolitical era if acquisitions involve exotic steels for state-of-the-art weapons systems. Focusing concern is the thought that, when other governments are involved even indirectly, takeovers represent "cross-border nationalization."[34]

Even when national security is not at issue, there remains the question of how to regard core industries coming under foreign ownership and control. There have been major foreign buyouts of American steelmakers, but three principal ones – United States Steel, Nucor, and AK Steel – remain independent. Those three, however, are regarded as fully in play. The same is not true in Canada, where all three steelmakers – Dofasco, Algoma, and Stelco – passed under foreign ownership in less than two years. There are two contrasting views of what constitutes the national interest.

The first equates sovereignty and independence with domestic ownership. Profits are retained at home, management is in domestic hands, and decisions regarding investment, employment, and output are made locally. Although industries cannot be made to adhere to public priorities without the kind of involvement found in Germany's tripartite representation of management, labour, and government interests (or, less directly, in France's earlier efforts with indicative planning), those industries do operate in a national context. Investment and market strategies are set in domestic boardrooms, not foreign ones. When plant closings and layoffs do occur, the decisions are made within the immediate sphere of national markets and constraints. If these firms invest abroad, the same managerial agency simply extends there. Although practices like these are not incompatible with trade, they can evolve into protectionism and subsidization. At their worst they support parochialism, inefficiency, and complacency. The national interest is defined in terms of authenticity and control.

The second view has its roots in liberal economics. Opening firms to international investment makes available larger stocks of capital and paves the way for investors with potentially wider horizons. International managers may see fuller opportunities for growth and development and draw on more ample troves of credit and expertise. Facilities that have static prospects at home may be reordered to enjoy more dynamic ones abroad. A presence in multiple commercial networks encourages economies of scope, and diverse assets can be organized for economies of scale. International managers may inject new vigour and discipline. At the same time, foreign owners have discretion over profits, output, and investment, and their priorities may favour venues elsewhere. Also, the viability of particular operations is determined with reference to a broader set of installations and markets, and cutbacks, layoffs, and closures are decided accordingly. Finally, foreign owners may not understand their new acquisitions and do a worse job of directing them than their domestic predecessors. Nonetheless, the prospect is economic efficiency and productivity, and these define the national interest.

International agreements and opinions are trending toward the liberal view.[35] States seeking to protect firms from foreign takeovers or to provide massive bailouts are viewed as exceptional. That point can be seen in the French government's efforts to block Mittal's takeover of Arcelor and the Swiss pharmaceutical firm Novartis' takeover of Aventis, and to save the engineering firm Alstom from bankruptcy with an infusion exceeding EU guidelines. Because these efforts cause difficulty with investment rules, other governments, and (in the Alstom case) the EU, they tend not to be the norm. Similarly, large bailouts, since they may not ensure a firm's long-term survival, are now regarded as dubious uses of public money. That point can be seen in the divided views in Canada and the United States in 2008 over bailing out the Big Three automakers. In contrast, interventions under competition law are widely viewed as legitimate and necessary. The weight of current policies and rules favours a continuation of international openness. The contrary force is national security and the uncertain consequences of investments backed by foreign governments. That force may strengthen if the Russian government's behaviour becomes dangerously adversarial or if China's economic and military expansion appears menacing.

Four Reasonable Prognostications

In light of all that has been seen, these four prognostications follow naturally:

1   World demand for steel will not decline in the near future. The reasons are steel's widespread utility and versatility, its diffusion in standardized technologies and products, and the homogeneity of its commodity grades. Despite concerns about automobile emissions and global warming, rapidly expanding markets in industrializing countries promise at least a steady demand for sheet steel. Efforts to develop and expand infrastructures promise the same for commodity steel.

2   Consolidation will deliver some of the expected advantages. Specialized facilities will become practical, not necessarily for entire global markets but certainly for sizable domestic ones. To the extent that occurs, additional scale economies will be possible, benefiting both prices and profitability. Acquired producers with established technological and market advantages will keep them because the incentives present in large and wealthy markets, which originally encouraged those advantages, are likely to remain. The acquired producers will be used as sources of expertise and experience to upgrade other installations in the firm.

3   Overinvestment will continue to be a problem because demand in rapidly growing economies is difficult to forecast and because individual incentives relating to market share and revenue may be stronger than the industry leaders' dissuasive counsel. A high level of consolidation in the industry would concentrate their message, but its influence would remain advisory. Over the long run, the risk is not of falling demand but of overinvestment and surpluses in the face of strong demand.

4   Blended global firms will reduce some of the informal costs of trade, but other costs will be more durable. Of these, transportation costs are the most important. They are high enough to limit other advantages, as was seen in Chapter 4 with respect to China. Given steel's weight and density, it will remain sensitive to transportation costs. If those costs again rise to early 2008 levels, they would limit the potential for global specialization, favour the present diversity of locales, encourage commodity-steel producers in growing markets to diversify (for which consolidated ownership would provide the needed technology), and constitute a further reason for buying resident producers. If transportation costs rise on higher fuel prices and remain there, the prospect is for a retraction of supply linkages in all industries where transportation costs are significant and where more proximate alternative sources are available.

Steel and the World Economy

What do developments in the steel industry imply more generally about the world economy? Five points stand out.

1   The importance of technology. It directly influences crucially important cost structures. These in turn affect economies of scale and pricing power as well as the ability to supply desired goods, to regulate production according to demand, and to withstand economic downturns. Standardized and portable technology enables industries to migrate to new markets and exploit either inherent trade advantages or those fostered by state intervention. It also enables new producers to bypass the time and costs of developing technology on their own and to progress beyond basic product levels. These factors bear directly on the international division of labour.

2   The importance of standardized goods. In both domestic and international markets, standardization eliminates qualitative barriers and affords new producers immediate entree. For producers lacking the resources to develop the kinds of innovations that change buyer preferences or create new markets, standardization is a strong equalizer.

3   The importance of economies of scale and reasonably priced distribution. These economic geography factors, to the extent they operate, make it

possible for producers to expand beyond local markets, become more efficient, and specialize. Trading costs may limit the force of these factors in distant markets; at the same time, they encourage industries to locate in places that simple comparative advantage would not select, particularly when the state intervenes to help secure initial and necessary factors. Examples of such locales in steel production are Japan, South Korea, and China. Whether such ventures succeed depends on their alignment with markets, the appropriateness of their technology, the ability of the firms to learn and adapt, and the nature of competition.

4   The importance of basic goods. Those with widespread utility, regardless of how prosaic they may seem, will continue to be demanded and to support investment and trade. That is especially true in rapidly developing economies where those goods are put to multiplying uses, but it is true more qualitatively in mature economies with specialized and sophisticated requirements. Both kinds of demand can be transformative. Only thirty years ago the steel industry was seen as mature and as edging into senility. Investment in new technology, an explosion of demand abroad, and development at the leading edges of metallurgy have bestowed new vigour and appeal.

5   The importance of being responsive to change. The established steel producers that survived the dismal 1960s and 1970s were the ones that were able to identify their areas of existing or prospective advantage and organize to meet them. That only a handful were able to do so independently shows how difficult those transitions can be, particularly when change is expensive, capital markets lose confidence, and advantages are difficult to discern. From a different perspective, that experience highlights the risks of complacency amidst changing technology and competition. The sources of changes expand with international trade. Although change occurs more slowly in industries such as steel than in more dynamic and open-ended sectors, steel's crisis decades show how disruptive change can be. Cyclicality multiplies the potential.

These implications apply more generally to industries in a world economy and can be stated concisely. The common elements are:

- larger, more open, and more complex markets
- an international mobility of goods and investment
- an ability of both industries and demand to migrate
- an ability of firms to combine and recombine
- an ability of technology to be installed and transferred over once-wide market and geographic distances.

These elements summarize the industrial aspects of globalization. Together they represent openness and fluidity across regions, states, and industries. Propelling movement are demand, incentives to organize in the face of market opportunities and cost constraints, and the ability to arrange resources. Globalization's effect is to magnify scale – of production, organization, markets, opportunities, and risk.

Notes

Chapter 1: Introduction

1 Roger S. Ahlbrandt, Richard J. Fruehan, and Frank Giarratani, *The Renaissance of American Steel: Lessons for Managers in Competitive Industries* (New York: Oxford University Press, 1996), 30.

2 Peter J.B. Steele, "The Propensity of Advanced Free World Economies to Import Steel," *Canada–United States Law Journal* 2 (1979): 20.

3 London Metal Exchange, "Steel Industry Overview," accessed March 28, 2007, at http://www.lme.com/steel-industryoverview.

4 IISI, *World Steel in Figures 2008* (Brussels: 2008), 6.

5 Even so, to assist Stelco's exit from bankruptcy, the Ontario government pledged $150 million to help fund Stelco's employee pension plan.

6 For an account of the rise of Mittal Steel and its takeover of Arcelor, see Tim Bouquet and Byron Ousey, *Cold Steel: Britain's Richest Man and the Multi-Billion-Dollar Battle for a Global Empire* (Boston: Little, Brown, 2008).

7 IISI, *World Steel in Figures 2008*, World Trade in Steel Products, 1975-2007, 20.

8 On trade-related investment measures generally, see Jeffrey S. Thomas and Michael Meyer, *The New Rules of Global Trade: A Guide to the World Trade Organization* (Scarborough: Carswell, 1997), 196-202.

9 "Panel Formed to Review Foreign Investment," *Northern Miner,* July 23-29, 2007, 5.

10 Deborah Solomon, "Foreign Investors Face New Hurdles Across the Globe: China, Canada, Russia Grow Wary of Acquirers," *Wall Street Journal,* July 6, 2007, A1.

11 The panel's report, "Compete to Win," can be accessed at http://www.competitionreview.ca.

12 Stephanie Kirchgaessner, "Tougher Scrutiny of Foreign Takeovers in US," *Financial Times,* July 27, 2007, 5.

13 Niels C. Sorrells and Andrew Peaple, "Politics and Economics: Germany Seeks EU Curbs on Some Foreign Takeovers," *Wall Street Journal*, July 19, 2007, A8. On existing German and EU rules, see GAO, *Foreign Investment: Laws and Policies Regulating Foreign Investment in 10 Countries,* Report 08-320 (Washington: February 2008).

14 David Nicholson, "Fortifying German Defences," *European Venture Capital and Private Equity Journal* 151 (April 2008): 8; Marcus Walker, "Germany Tinkers with Foreign-Takeovers Plan," *Wall Street Journal,* January 14, 2008, A2.

15 James Simms, "Japan's Merger Angst: New Law Won't Slow Deals," *Wall Street Journal,* April 10, 2007, C10.

16 Donald R. Davis and David E. Weinstein, "Economic Geography and Regional Production Structure: An Empirical Investigation," *European Economic Review* 43 (February 1999): 402-3.

17 Mitsuo Matsushita, Thomas J. Schoenbaum, and Petros C. Mavroidis, *The World Trade Organization: Law, Practice, and Policy* (Oxford: Oxford University Press, 2003), 278.

18 Ibid., 293, 299.
19 Bernard M. Hoekman and Michel M. Kostecki, *The Political Economy of the World Trading System: The WTO and Beyond* (Oxford: Oxford University Press, 2001), 331.
20 Ibid., 312-13.
21 For a timeless observation about the way that being a mathematician affects one's understanding of contests and strategies, see Thomas C. Schelling, *The Strategy of Conflict* (New York: Oxford University Press, 1963), 113-14.
22 William H. Branson and Alvin K. Klevorick, "Strategic Behaviour and Trade Policy," in *Strategic Trade Policy and the New International Economics,* ed. Paul R. Krugman (Cambridge, MA: MIT Press, 1986), 251-52.
23 Peter Marsh, "Mittal Puts a Shine on European Steel," *Financial Times,* April 19, 2007, 25.
24 Ahlbrandt et al., *The Renaissance of American Steel,* focuses on management strategies. Anthony D'Costa, *The Global Restructuring of the Steel Industry: Innovations, Institutions, and Industrial Change* (New York: Routledge, 1999), examines the industry's transformation, but it was written before the industry's shakeout in 2000-1, before the surging growth of steel demand and production in China, and before the current trend toward large-scale consolidation.

Chapter 2: A Tough Industry

1 Roger S. Ahlbrandt, Richard J. Fruehan, and Frank Giarratani, *The Renaissance of American Steel: Lessons for Managers in Competitive Industries* (New York: Oxford University Press, 1996), 34-37.
2 Concise and clearly presented information on steel technology and metallurgy can be found on the website of the American Iron and Steel Institute at http://www.steel.org.
3 Ahlbrandt et al., *The Renaissance of American Steel,* 36.
4 Steve Karpel, "Resurgent Iron Age," *Metal Bulletin Monthly,* March 2006, 16.
5 For process descriptions of direct reduction and Corex, see A.F. Eberle, W. Kepplinger, and S. Zeller, "VAI Technologies for Scrap Substitutes," *The Steel Industry in the New Millennium, Vol. I: Technology and the Market,* ed. Ruggero Ranieri and Jonathan Aylen (London: IOM Communications, 1998), 117-18. This collection of engineering and economics papers, presented at an international conference in Terni, Italy, in 1996, is an excellent survey of steelmaking technology.
6 Wallace Huskonen, "Ironmaking by Another Route," *Metal Producing and Processing,* January-February 2006, 15. The article includes a process diagram of the technology.
7 F. Marcus, "Ironies in Steel," in *The Steel Industry in the New Millennium,* ed. Ranieri and Aylen, 17-25.
8 W. Hogan, "Prospects for the Steel Industry into the New Millennium," in *The Steel Industry in the New Millennium,* ed. Ranieri and Aylen, 323.
9 F. Bonomo, "Summary: Technological Innovation," in *The Steel Industry in the New Millennium,* ed. Ranieri and Aylen, 112; H.B. Lungen, "Technological Innovations in Iron and Steelmaking and their Effects on Coal, Coke, and Iron Ores," in *The Steel Industry in the New Millennium,* ed. Ranieri and Aylen, 43-57.
10 IISI, *World Steel in Figures 2008,* Crude Steel Production by Process, 2007 (Brussels: 2008), 9.
11 Anthony Poole, "Chinese Steel Demand Growth to Slow for Remainder of Decade," *Platt's International Coal Report,* March 27, 2006, 11.
12 IISI, *World Steel in Figures 2008,* 9; Karpel, "Resurgent Iron Age," 16.
13 Paul Glader, "China's Baosteel Looks Abroad," *Wall Street Journal,* March 31, 2004, A6.
14 Diana Kinch, "Latin America's Costs Deter," *Metal Bulletin Monthly,* April 2006, 10.
15 Diana Kinch, "Latin America: Gerdau Springs Forward," *Metal Bulletin Monthly,* May 2006, 10.
16 Jacques E. Astier, "Evolution of the World Iron Ore Market," *Minerals and Energy* 16 (December 2001): 26.
17 Jonathan Aylen, "Trends in the International Steel Market," in *The Steel Industry in the New Millennium,* ed. Ranieri and Aylen, 212.
18 E. Biamonte and A.B. Hinder, "The Engineering Business in the Early Decades of the Next Century," in *The Steel Industry in the New Millennium,* ed. Ranieri and Aylen, 87.

19 Kamachi Ruthramurthi-Nillsen, "Planning for a Boom," *Metal Bulletin Monthly*, May 2006, 27.
20 Corus Steel's website provides well-illustrated information about the steelmaking process at http://www.coruseducation.com.
21 Douglas Alan Fisher, *Steel: From the Iron Age to the Space Age* (New York: Harper and Row, 1967), 17-18.
22 Jang-Sup Shin, *The Economics of the Latecomers: Catching Up, Technology Transfer, and Institutions in Germany, Japan, and South Korea* (London: Routledge, 1996), 68.
23 Jon R. Neill, "Production and Production Functions: Some Implications of a Refinement to Process Analysis," *Journal of Economic Behaviour and Organization* 51 (August 2003): 507-21, presents a production decision analysis of the open-hearth process.
24 Bryan Berry, "A Restrospective on Twentieth-Century Steel – Technology," *New Steel*, November 1999.
25 G.S. Maddala and Peter T. Knight, "International Diffusion of Technical Change: A Case Study of the Oxygen Steelmaking Process," *Economic Journal* 77 (September 1967): 538-39.
26 Berry, "A Retrospective."
27 Hajime Sato, "Total Factor Productivity vs. 'Realism': The Case of the South Korean Steel Industry," *Cambridge Journal of Economics* 29 (July 2005): 651.
28 H.G. Baumann, "The Relative Competitiveness of the Canadian and U.S. Steel Industries, 1955-1970," *Economia Internazionale* 27 (February 1974): 154.
29 Berry, "A Retrospective."
30 Shin, *The Economics of the Latecomers*, 95-96.
31 IISI, *World Steel in Figures 2008*, 9.
32 Eberle et al., "VAI Technologies for Scrap Substitutes," 114.
33 Ibid., 113.
34 Berry, "A Retrospective."
35 E. Repetto, G. Brascugli, and G. Perni, "Evolution of EAF Steelmaking Route," in *The Steel Industry in the New Millennium*, ed. Ranieri and Aylen, 137.
36 Michael Valenti, "Vacuum Degassing Yields Stronger Steel," *Mechanical Engineering*, April 1998, 54-58, provides a clear explanation of the chemistry and technology.
37 An excellent and illustrated description of rolling processes can be found in William T. Lankford, Jr., et al., eds., *The Making, Shaping, and Treating of Steel*, 10th ed. (Pittsburgh: Association for Iron and Steel Technology, 1985). An index of thoroughness is the book's 1,572-page length.
38 John C. McKay, "History, Iron and Steel Industry," *Canadian Encyclopedia*, accessed on August 26, 2008 at http://www.thecanadianencyclopedia.com; personal communication, August 18, 2008.
39 Baumann, "The Relative Competitiveness of the Canadian and U.S. Steel Industries," 154.
40 Shin, *The Economics of the Latecomers*, 96.
41 IISI, *World Steel in Figures 2008*, Continuously Cast Steel Output, 2005-2007, 16.
42 Robert R. Miller, "The Changing Economics of Steel," *Finance and Development* 28 (June 1991): 39.
43 Aylen, "Trends in the International Steel Market," 203, 211.
44 Berry, "A Retrospective."
45 John Sheridan, "More Steel Productivity Gains Ahead?" *Industry Week*, August 15, 1997, 86-92.
46 Executive roundtable, "The Changing Dynamics of Steelmaking," *New Steel*, August 1999.
47 Peter Warrian and Celine Mulhern, "Knowledge and Innovation in the Interface Between the Steel and Automotive Industries: The Case of Dofasco," *Regional Studies* 39 (April 2005): 162.
48 Marcus, "Ironies in Steel," 16.
49 IC, *Primary Steel in Canada: Industry Snapshot* (Ottawa: 2000), 9.
50 B. O'hUallachain, "The Restructuring of the U.S. Steel Industry: Changes in the Location of Production and Employment," *Environment and Planning A* 26 (September 1993): 1339.

51 Margaret Hunt, "Dynamic Steel Industry," *Advanced Materials and Processes*, January 1998, 4.
52 Kenneth Warren, *World Steel: An Economic Geography* (New York: Crane, Russak, 1975), 95. These figures are updated to reflect current mill capacities.
53 "Essar Steel Consolidates Presence in N. America," *Businessline*, May 3, 2008.
54 Astier, "Evolution of the World Iron Ore Market," 28.
55 Warren, *World Steel*, 28-29.
56 Sandra Buchanan, "Fighting Fit," *Metal Bulletin Monthly*, May 2006, 19.
57 Andy Blamey, "EC Asks for Input on Future Policy for EU Metals Industry," *Platt's Metals Week*, September 25, 2006, 12.
58 James A. Schmitz, Jr., "What Determines Productivity? Lessons from the Dramatic Recovery of the U.S. and Canadian Iron Industries Following Their Early 1980s Crisis," *Journal of Political Economy* 113 (June 2005): 582-625.
59 Kevin Foster, "Asia: Chinese Weight," *Metal Bulletin Monthly*, April 2006, 10.
60 "Forging a New Shape," *The Economist*, December 10, 2005, 68; "China Agrees to Pay More for Iron Ore Imports," *International Herald Tribune*, accessed on June 23, 2008, at http://www.iht.com.
61 Tim Johnston, "Rio Tinto Balks at Benchmark Ore Price Deal," *International Herald Tribune*, February 21, 2008, accessed online on June 23, 2008, at http://www.iht.com.
62 Robert Guy Matthews, "Rio Tinto Cuts Iron Output to Stop Freefall," *Wall Street Journal*, November 11, 2008, B2.
63 Peter Marsh, "Asian Steelmakers' Deal Fuels Concern," *Financial Times*, February 19, 2008, 17; Thomas Gryta, "U.S. Steel May Continue to Shine," *Financial Post*, April 2, 2007, FP8.
64 M. Kipping, "Cooperation Between Steel Producers and Steel Users: A Major Determinant of National Competitive Advantage," in *The Steel Industry in the New Millennium*, ed. Ranieri and Aylen, 219.
65 Alexi Mordashov, "Why My Money Is on a Global Expansion in Steel," *Financial Times*, June 16, 2006, 15.
66 "Mittal Growing Again with Ukraine Acquisitions," *Metal Producing and Processing*, November-December 2005, 10.
67 Matthew Boyle, "2006 Thermal Coal Settlement Delay Stalls Semi-Soft Coking Coal, PCI Contract Negotiations," *Platt's International Coal Report*, April 3, 2006, 1.
68 Neil Hume, Julie MacIntosh, and Peter Marsh, "High Cost of Iron Ore and Coal Hurts Steel Producers," *Financial Times*, June 16, 2008, 15; ArcelorMittal, Annual Report 2007, 6.
69 Executive roundtable, "The Changing Dynamics of Steelmaking."
70 Ahlbrandt et al., *The Renaissance of American Steel*, 24.
71 O'hUallachain, "The Restructuring of the U.S. Steel Industry," 1345.
72 Ahlbrandt et al., *The Renaissance of American Steel*, 24.
73 IISI, *World Steel in Figures 2008*, Scrap, Estimated Consumption, Trade, and Apparent Domestic Supply, 2007, 23.
74 Anthony D'Costa, *The Global Restructuring of the Steel Industry: Innovations, Institutions, and Industrial Change* (New York: Routledge, 1999), 33.
75 IC, *Primary Steel: Overview and Prospects* (Ottawa: 1996), 37.
76 Stephen Power, "Take It Back: Where Do Cars Go When They Die?" *Wall Street Journal*, April 17, 2006, R6; Jarret Bilous and Kam Hon, *North American Steel Industry Study* (Toronto: Dominion Bond Rating Service, March 2004), 14.
77 "Costs," accessed on July 22, 2008, at http://www.steelonthenet.com; Robert Guy Matthews, "Metals Meltdown Burns Scrap Dealers," *Wall Street Journal*, October 24, 2008, B1.
78 Miller, "The Changing Economics of Steel," 39.
79 IISI, *World Steel in Figures 2008*, Crude Steel Production by Process 2007, 9.
80 Ahlbrandt et al., *The Renaissance of American Steel*, 26.
81 Peter Clancy, *Micropolitics and Canadian Business: Paper, Steel, and the Airlines* (Peterborough: Broadview, 2004), 43, 197.
82 IC, *Primary Steel: Overview and Prospects*, 20.
83 "Made in the U.S. – But Owned in Europe?" *Business Week*, February 19, 2001, 92.
84 Michael Moore, "Waning Influence of Big Steel?" in *The Political Economy of American Trade Policy*, ed. Anne O. Krueger (Chicago: University of Chicago Press, 1996), 78.

85 Jose Guilherme de Heraclito Lima, *Restructuring the U.S. Steel Industry: Semi-Finished Steel Imports, International Integration, and U.S. Adaptation* (Boulder: Westview, 1991), 35.
86 Stephen Woolcock, "Iron and Steel," in *The International Politics of Surplus Capacity,* ed. Susan Strange and Roger Tooze (London: George Allen and Unwin, 1981), 78.
87 *Value Line Investment Survey,* April 25, 2008, 1414.
88 Ibid., 1410.
89 Ibid.
90 *Value Line Investment Survey,* October 24, 2008, 1414.
91 Aaron Tornell, "Rational Atrophy: The U.S. Steel Industry," Working Paper 6084 (Cambridge, MA: National Bureau of Economic Research, 1997), 13.
92 Michel Glais, "Steel Industry," in *European Policies on Competition, Trade, and Industry: Conflict and Complementarities,* ed. Pierre Buiges, Alexis Jacquemin, and Andre Sapir (Aldershot: Elgar, 1995), 220.
93 Heraclito Lima, *Restructuring the U.S. Steel Industry,* 35.
94 Michael Moore, "Made in America? U.S. Steelmaking in the 1990s: An American Resurgence?" *International Executive* 38 (July-August 1996), 442.
95 Donald F. Barnett and Louis Schorsch, *Steel: Upheaval in a Basic Industry* (Cambridge, MA: Ballinger, 1983), 179-80.
96 Baumann, "The Relative Competitiveness of the Canadian and U.S. Steel Industries," 150.
97 Bernhard Fischer, Juan-Carlos Herken-Krauer, Matthias Lucke, and Peter Nunnenkamp, *Capital-Intensive Industries in Newly Industrializing Countries: The Case of the Brazilian Automobile and Steel Industries* (Tübingen: Mohr Siebeck, 1988), 214.
98 David G. Tarr, "The Minimum Efficient Size Steel Plant," *Atlantic Economic Journal* 12 (March 1984): 122.
99 Glais, "Steel Industry," 221; Heraclito Lima, *Restructuring the U.S. Steel Industry,* 35.
100 George W. Stocking, *Basing Point Pricing and Regional Development: A Case Study of the Iron and Steel Industry* (Chapel Hill: University of North Carolina Press, 1954), 25.
101 Robert Guy Matthews, "US Steel Triples Its Net Income," *Wall Street Journal,* October 29, 2008, B2.
102 OECD Directorate for Financial, Fiscal, and Enterprise Affairs, Committee on Competition Law and Policy, *Oligopoly* (Paris: October 19, 1999), 27.
103 Peter Marsh, "Mittal Fatigue," *Financial Times,* October 31, 2008, 8; Myrna Pinkham, "North America: What Goes Up," *Metal Bulletin Monthly,* November 2008, 10.
104 Heraclito Lima, *Restructuring the U.S. Steel Industry,* 34.
105 Industry Canada, *Primary Steel: Overview and Prospects,* 18.
106 Ahlbrandt et al., *The Renaissance of American Steel,* 26.
107 Peter J.B. Steele, "The Propensity of Advanced Free World Economies to Import Steel," *Canada–United States Law Journal* 2 (1979): 18.
108 William Kilbourn, *The Elements Combined: A History of the Steel Company of Canada* (Toronto: Clarke, Irwin, 1960), 230.
109 Glais, "Steel Industry," 220; Richard G. Harris, "Trade and Industrial Policy for a 'Declining' Industry: The Case of the U.S. Steel Industry," in *Empirical Studies of Strategic Trade Policy,* ed. Paul R. Krugman and Alasdair Smith (Chicago: University of Chicago Press, 1984), 140.
110 IISI, *World Steel in Figures 2008,* Top Steel-Producing Companies in 2006 and 2007, 7.
111 Ibid.
112 For a set of varying estimates, see Fischer et al., *Capital-Intensive Industries in Newly Industrializing Countries,* 220.
113 IISI, *World Steel in Figures 2008,* Top Steel-Producing Companies, 2006 and 2007, 7.
114 Donald F. Barnett and Robert W. Crandall, *Up from the Ashes: The Rise of the Steel Minimill in the United States* (Washington: Brookings Institution, 1986), 3.
115 Pankaj Ghemawat, *Games Businesses Play: Cases and Models* (Cambridge, MA: MIT Press, 1997), 154.
116 Donald F. Barnett and Robert W. Crandall, "Steel: Decline and Renewal," in *Industry Studies,* 2nd ed., ed. Larry L. Deutsch (Armonk, NY: M.E. Sharpe, 1998), 128.
117 Bernard Keeling, "Structural Change in the World Steel Industry: A North-South Perspective," in *Industry on the Move: Causes and Consequences of Industrial Location in the Manufacturing*

Industry, ed. Gijsbert van Liemt (Geneva: International Labour Organization, 1992), 162; Ahlbrandt et al., *The Renaissance of American Steel*, 132.

118 OECD, *Oligopoly*, 27.
119 Baumann, "The Relative Competitiveness of the Canadian and U.S. Steel Industries," 148.
120 M. Kipping, "Co-Operation Between Steel Producers and Steel Users," 215, 230.
121 Sato, "Total Factor Productivity vs. 'Realism,'" 651-52.
122 O' hUallachain, "The Restructuring of the U.S. Steel Industry," 1340.
123 Clancy, *Micropolitics and Canadian Business*, 51.
124 Moore, "Made in America?" 441.
125 Richard E. Caves and Michael E. Porter, "Barriers to Exit," in *Essays on Industrial Organization in Honor of Joe S. Bain*, ed. Robert T. Masson and P. David Qualls (Cambridge, MA: Ballinger, 1976), 69.
126 Michael J. Hiscox, "Commerce, Coalitions, and Factor Mobility: Evidence from Congressional Votes on Trade Legislation," *American Political Science Review* 96 (September 2002): 597.
127 Caves and Porter, "Barriers to Exit," 39.
128 Ibid., 49-50.
129 Steele, "The Propensity of Advanced Free World Economies to Import Steel," 18.
130 Woolcock, "Iron and Steel," 75.
131 Stocking, *Basing Point Pricing and Regional Development*, 26.
132 OECD, *Oligopoly*, 27.
133 Moonjoong Tcha and Larry A. Sjaastad, "Analysis of Steel Prices," in *The Economics of the East Asia Steel Industries*, ed. Yanrui Wu (Aldershot: Ashgate, 1998), 208.
134 Warrian and Mulhern, "Knowledge and Innovation," 162.
135 Berry, "A Retrospective."
136 Industry Canada, *Primary Steel in Canada*, 5.
137 Ibid., 14; Tcha and Sjaastad, "Analysis of Steel Prices," 207.
138 Joachim Oliveira Martins, "Market Structure, Trade, and Industry Wages," *OECD Economic Studies* 22 (Paris: Spring 1994).
139 Ghemawat, *Games Businesses Play*, 153.
140 Warren, *World Steel*, 91.
141 Roger Philips, Management Roundtable, "The Changing Dynamics of Steelmaking," *New Steel*, August 1999.
142 Gary Clyde Hufbauer and Ben Goodrich, "Steel: Big Problems, Better Solutions," Policy Brief 01-9 (Washington: Institute for International Economics, July 2001), accessed online on October 1, 2006, at http://www.iie.com/publications/pb/bp.cfm?ResearchID=77.
143 Robert D. Willig, "Economic Effects of Antidumping Policy," in Robert Z. Lawrence, ed., *Brookings Trade Forum 1998* (Washington: Brookings Institution, 1998), 63.
144 Hufbauer and Goodrich, "Steel: Big Problems, Better Solutions."
145 OECD, *Oligopoly*, 27.
146 Moore, "Waning Influence of Big Steel?" 78.
147 Barnett and Schorsch, *Steel*, 220.
148 Baumann, "The Relative Competitiveness of the Canadian and U.S. Steel Industries," 150.
149 Harris, "Trade and Industrial Policy for a Declining Industry," 139; Barnett and Crandall, "Steel: Decline and Renewal," 135.
150 OECD, *Oligopoly*, 24.
151 Barnett and Crandall, "Steel: Decline and Renewal," 135.
152 M. Kipping, "Cooperation Between Steel Producers and Steel Users," 219.
153 Caves and Porter, "Barriers to Exit," 46-48.
154 Glais, "Steel Industry," 221.
155 Barnett and Crandall, "Steel: Decline and Renewal," 135.
156 Tonya Vinas, "Steel Prices Rise – Again," *Industry Week*, March 2004, 12.
157 2008 Steel Prices, on July 18, 2008 at http://steelonthenet.com. As was seen earlier, insulation from world ore prices is an advantage of North American steelmakers.
158 Glais, "Steel Industry," 221.
159 GAO, *International Trade: The Health of the U.S. Steel Industry*, Report no. B-236037 (Washington: July 1989), 34.

160 *Iron and Steel Mills 1960* (Ottawa: Dominon Bureau of Statistics, 1963), 6; *Steel Mills 1971* (Ottawa: Statistics Canada, 1973), 4; *Primary Steel, Steel Pipe, and Tube Industries and Iron Foundries, 1984* (Ottawa: Statistics Canada, 1986), 8.
161 D'Costa, *The Global Restructuring of the Steel Industry*, 16-17.
162 Industry Canada, *Primary Steel: Overview and Prospects*, 35.
163 Shin, *The Economics of the Latecomers*, 96-97.
164 Michael Verespej, "Steel's Dilemma," *Industry Week*, July 16, 2001, 3; Bilous and Hom, *North American Steel Industry Study*, 20.
165 Keeling, "Structural Change in the World Steel Industry," 163.
166 Warren, *World Steel*, 79.
167 Ibid., 85. A checklist of reasons for and against modernizing can be found in Bela Gold, William S. Pierce, Gerhard Rosegger, and Mark Perlman, *Technological Progress and Industrial Leadership: The Growth of the U.S. Steel Industry, 1900-1970* (Lexington: Lexington, 1984), 58, 60. The decisional process is discussed in that book on pages 60-63.
168 D'Costa, *The Global Restructuring of the Steel Industry*, 45.
169 W.L. Dack, "Canada's Steel Industry Expands in a Big Way," *Canadian Geographical Journal* 91 (October 1975): 37.
170 Hans G. Mueller, "The Steel Industry," *Annals of the American Academy of Political and Social Science* 460 (March 1982): 75.
171 Keeling, "Structural Change in the World Steel Industry," 163; D'Costa, *The Global Restructuring of the Steel Industry*, 18.
172 OECD Business and Industry Policy Forum, Forum on Global Industrial Restructuring, Report: "Recent Trends in Global Industrial Restructuring" (Paris: February 19, 2002), 9.
173 "Pick and Mix," *Metal Bulletin Monthly*, November 2005, 9.
174 Barnett and Schorsch, *Steel*, 204.
175 "Pick and Mix," 9.
176 Industry Canada, *Primary Steel: Overview and Prospects*, 30; Jim Forbes, "Merger Myopia," *Metal Bulletin Monthly*, November 2005, 16.
177 Ibid.
178 C. Sherman Cheung, Itzhak Krinsky, and Bernadette Lynn, "The Canadian Steel Industry: Current State and Future Prospects," *Canadian Banker* 92 (April 1985): 8.
179 Baumann, "The Relative Competitiveness of the Canadian and U.S. Steel Industries," 154.
180 Dack, "Canada's Steel Industry Expands in a Big Way," 34.
181 Barnett and Schorsch, *Steel*, 218-20.
182 Anthony Masi, "Structural Adjustment and Technological Change in the Canadian Steel Industry, 1970-1986," in *The New Era of Global Competition: State Policy and Market Power*, ed. Daniel Drache and Meric S. Gertler (Montreal and Kingston: McGill-Queen's University Press, 1991), 192.
183 M.J. Webber, "Regional Production and the Production of Regions: The Case of Steeltown," in *Production, Work, Territory: The Geographical Anatomy of Industrial Capitalism*, ed. Allan J. Scott and Michael Storper (Boston: Allen and Unwin, 1986), 217-19.
184 Masi, "Structural Adjustment and Technological Change," 191.
185 Barnett and Crandall, *Up from the Ashes*, 107.
186 Kent Jones, "Structural Adjustment in the United States Steel Industry," in *Industry on the Move: Causes and Consequences of International Relocation in the Manufacturing Industry*, ed. Gijsbert van Leimt (Geneva: International Labour Organization, 1992), 187.
187 Tornell, "Rational Atrophy," 13-14.
188 Barnett and Schorsch, *Steel*, 225-26.
189 Ibid.
190 Adam Ritt, "The Danger of Devaluation," *New Steel*, September 1999.
191 Bilous and Hon, *North American Steel Industry Study*, 7.
192 IC, *Primary Steel in Canada: An Industry Snapshot*, 5.
193 IC, *Primary Steel: Overview and Prospects*, 28-30.
194 Ibid., 41-42.
195 IC, *Primary Steel in Canada: Industry Snapshot*, 5.
196 Kris Inwood, "The Iron and Steel Industry," in *Progress Without Planning: The Economic His-*

tory of Ontario from Confederation to the Second World War, ed. Ian M. Drummond (Toronto: University of Toronto Press, 1987), 199-204.

197   A review of events appears in Thomas Watson, "What a Mess," *Canadian Business,* January 30, 2006, 36-44.

198   Thomas W. Gerdel, "Steel Industry's Ills Taking Toll in Canada," *Cleveland Plain Dealer,* February 19, 2004.

199   Masi, "Structural Adjustment and Technological Change," 192.

200   IC, *Primary Steel: Overview and Prospects,* 1.

201   Nelson Schwartz, "Bent but Unbowed," *Fortune,* July 22, 2002, 118.

202   Ahlbrandt et al., *The Renaissance of American Steel,* 24; William R. Cline, "U.S. Strategic Trade and Industrial Policy: The Experience of Textiles, Steel, and Automobiles," in *Strategic Trade Policy and the New International Economics,* ed. Paul R. Krugman (Cambridge, MA: MIT Press, 1996), 220.

203   Michael Moore, "Steel Protection in the 1980s: The Waning Influence of Big Steel?" Paper no. 4760 (Cambridge, MA: National Bureau of Economic Research, June 1994), 40; IISI, *World Steel in Figures 2008,* Continuously Cast Steel Output, 2005 to 2007, 16.

204   Hunt, "Dynamic Steel Industry," 4.

205   Ahlbrandt et al., *The Renaissance of American Steel,* 27, 58.

206   Mark Tatge, "Rusty Steelmakers Take a Shine," *Forbes,* July 5, 2004, 150.

207   Ahlbrandt et al., *The Renaissance of American Steel,* 27, 58.

208   Bilous and Hom, *North American Steel Study,* 23.

209   "Smeltdown," *The Economist,* October 20, 2001, 62-65.

210   Jones, "Structural Adjustment," 197-98.

211   "American Crossroads," *Metal Bulletin Monthly,* August 2005, 7.

212   "Big Is Back," *The Economist,* October 30, 2004, 67.

213   OECD, "Trends in Global Industrial Restructuring," 25.

214   Michael Arndt, "Melting Away Steel's Costs," *Business Week,* November 8, 2004, 48.

215   "Steeling for a Fight," *Time Atlantic,* March 18, 2002, 38.

216   James May, "Canadian Recovery," *Metal Bulletin Monthly,* August 2005, 16.

Chapter 3: Prices, Preferences, and Strategy

1   James Alt, Jeffrey Frieden, Michael Gilligan, Dani Rodrik, and Ronald Rogowski, "The Political Economy of International Trade: Enduring Puzzles and an Agenda for Inquiry," *Comparative Political Studies* 29 (December 1996): 692-94.

2   Robert Gilpin, *Global Political Economy: Understanding the Global Economic Order* (Princeton: Princeton University Press, 2001), 133.

3   Louis Philips, *The Economics of Price Discrimination* (Cambridge: Cambridge University Press, 1983), 23.

4   Paul Klemperer and Margaret Meyer, "Price Competition vs. Quantity Competition: The Role of Uncertainty," *Rand Journal of Economics* 17 (Winter 1986): 634.

5   Brink Lindsey, "The U.S. Antidumping Law: Rhetoric versus Reality," *Journal of World Trade* 34 (January 2000): 23.

6   George W. Stocking, *Basing Point Pricing and Regional Development: A Case Study of the Iron and Steel Industry* (Chapel Hill: University of North Carolina Press, 1954), 36.

7   OECD Directorate for Financial, Fiscal and Enterprise Affairs, Committee on Competition Law and Policy, *Oligopoly* (Paris: October 19, 1999), 20.

8   Ray Rees, "Tacit Collusion," *Oxford Review of Economic Policy* 9 (Summer 1993): 36.

9   Ibid., 35-36.

10   OECD, *Oligopoly,* 7.

11   Rees, "Tacit Collusion," 31.

12   Stephen Woolcock, "Iron and Steel," in *The International Politics of Surplus Capacity,* ed. Susan Strange and Roger Tooze (London: George Allen and Unwin, 1981), 73.

13   Carl Shapiro, "Theories of Oligopoly Behaviour," in *Handbook of Industrial Organization,* Vol. I, ed. Richard Schmalensee and Robert D. Willig (Amsterdam: North-Holland, 1989), 334.

14 Stocking, *Basing Point Pricing and Regional Development*, 39.
15 F.M. Scherer and David Ross, *Industrial Market Structure and Economic Performance*, 3rd ed. (Boston: Houghton Mifflin, 1990), 293. Quoted in OECD, *Oligopoly*, 44.
16 Paul R. Krugman, "Where Is the 'New Economic Geography'?" in *The Oxford Handbook of Economic Geography*, ed. Gordon L. Clark, Maryann Feldman, and Meric S. Gertler (Oxford: Oxford University Press, 2000), 52.
17 Rees, "Tacit Collusion," 39.
18 OECD, *Oligopoly*, 20.
19 N. Gregory Mankiw and Michael D. Whinston, "Free Entry and Social Inefficiency," *Rand Journal of Economics 17* (Spring 1986): 57.
20 Donald F. Barnett and Robert W. Crandall, *Up from the Ashes: The Rise of the Steel Minimill in the United States* (Washington: Brookings Institution, 1986), 13-14.
21 Hal R. Varian, "Price Discrimination," in *Handbook of Industrial Organization, Vol. I*, ed. Richard Schmalensee and Robert D. Willig (Amsterdam: North-Holland), 604.
22 Ibid., 646.
23 Melvin L. Greenhut, George Norman, and Chao-Shun Hung, *The Economics of Imperfect Competition* (Cambridge: Cambridge University Press, 1987), 160-61.
24 James A. Brander and Paul R. Krugman, "A 'Reciprocal Dumping' Model of International Trade," Discussion Paper #513 (Kingston: Queen's University, Institute for Economic Research, n.d.), 1-2.
25 Philips, *The Economics of Price Discrimination*, 16.
26 Rudnar Brannlund and Karl-Gustaf Onfgren, "Cyclical Dumping and Correlated Business Cycles in Imperfect Markets: Empirical Applications to the Canadian Pulp and Paper Industry," *Applied Economics* 27 (1995): 1092; Shapiro, "Theories of Oligopoly Behaviour," 343-45, 349.
27 Kenneth H. Kelly and Morris E. Morkre, "Do Unfairly Traded Imports Injure Domestic Industries?" *Review of International Economics* 6 (May 1998): 323.
28 Peter J.B. Steele, "The Propensity of Advanced Free World Economies to Import Steel," *Canada–United States Law Review* 2 (1979): 20, 23.
29 Susan Hutton and Michael Trebilcock, "An Empirical Study of the Application of Canadian Anti-Dumping Laws: A Search for Normative Rationales," *Journal of World Trade* 24 (June 1990): 127.
30 Wilfrid Ethier, "Dumping," *Journal of Political Economy* 90 (December 1982): 492.
31 C. Sherman Cheung, Itzhak Krinsky, and Bernadette Lynn, "The Canadian Steel Industry: Current State and Future Prospects," *Canadian Banker* 92 (April 1985): 10.
32 Steele, "The Propensity of Advanced Free World Economies to Import Steel," 18.
33 Gunnar Niels, "What Is Antidumping Policy Really About?" *Journal of Economic Surveys* 14 (September 2000): 472-73.
34 Jacob Viner, *Dumping: A Problem in International Trade* (New York: August M. Kelly, 1966), 139-41.
35 Ibid., 23.
36 Ibid., 138-43.
37 Christian A. Conrad, "Dumping and Anti-Dumping Measures from a Competition and Allocation Perspective," *Journal of World Trade* 36 (June 2002): 564.
38 Judith Goldstein, "The Political Economy of Trade: Institutions of Protection," *American Political Science Review 80* (March 1986): 165-66.
39 Niels, "What is Antidumping Policy Really About?" 475.
40 Ibid.
41 Lindsey, "The U.S. Antidumping Law," 20.
42 Phaedon Nicolaides, "The Competition Effects of Dumping," *Journal of World Trade 24* (October 1990): 119.
43 Ibid.
44 Stephen W. Davies and Anthony J. McGuinness, "Dumping at Less Than Marginal Cost," *Journal of International Economics* 21 (January 1982): 174.
45 An excellent decision matrix can be found in Hayne E. Leland, "Theory of the Firm Facing Uncertain Demand," *American Economic Review* 62 (1972): 279. For an elaboration, see Chin

Lim, "The Ranking of Behavioural Modes of the Firm Facing Uncertain Demand," *American Economic Review* 70 (1980): 217-24.

46 Nicolaides, "The Competition Effects of Dumping," 119.

47 Viner, *Dumping*, 142.

48 Lindsey, "The U.S. Antidumping Law," 21.

49 Richard Cooper, "Comments on Alan O. Sykes, 'Antidumping and Antitrust: What Problems Does Each Address?'" in *Brookings Trade Forum 1998*, ed. Robert Z. Lawrence (Washington: Brookings Institution, 1998), 47.

50 Robert W. Staiger and Frank Wolak, "The Effect of Domestic Antidumping Law in the Presence of Foreign Monopoly," *Journal of International Economics* 32 (May 1992): 267-69.

51 Robert D. Willig, "Economic Effects of Antidumping Policy," in *Brookings Trade Forum 1998*, 63; Joseph Stiglitz, "Dumping on Free Trade: The U.S. Import Trade Laws," *Southern Economic Journal* 64 (October 1997): 410.

52 Willig, "Economic Effects of Antidumping Policy," 66-67.

53 Nicolaides, "The Competition Effects of Dumping," 120.

54 David Weinstein, "Competition and Unilateral Dumping," *Journal of International Economics* 32 (May 1992): 387.

55 Thusnelda Tivig and Uwe Waltz, "Market Share, Cost-Based Dumping, and Anti-Dumping Policy," *Canadian Journal of Economics* 33 (February 2000): 70-71.

56 Willig, "Economic Effects of Antidumping Policy," 77. For a variation, see Donghyun Park, "Why Do Firms Dump at a Loss? An Economies-of-Scale Explanation," *Open Economies Review* 9 (1998): 258.

57 Richard Clarida, "Entry, Dumping, and Shakeout," *American Economic Review* 83 (March 1993): 181.

58 Willig, "Economic Effects of Antidumping Policy," 65.

59 Jonathan Eaton and Edward Mirman, "Predatory Dumping and Signal Jamming," in *Trade, Policy, and International Adjustments*, ed. Akira Takayama, Hiroshi Ohta, and Michihiro Ohyama (San Diego: Academic Press, 1991), 60-76.

60 Michael Moore, "Made in America? U.S. Steelmaking in the 1990s: An American Resurgence?" *International Executive* 38 (July-August 1996): 445.

61 James A. Brander, "Rationales for Strategic Trade and Industrial Policy," in *Strategic Trade Policy and the New International Economics*, ed. Paul R. Krugman (Cambridge, MA: MIT Press, 1986), 23-46.

62 Moore, "Made in America?" 445.

63 Peter Brieger, "Report Blames Subsidies for Glut of Steel," *Financial Post*, February 17, 2003.

64 Anthony D'Costa, *The Global Restructuring of the Steel Industry: Innovations, Institutions, and Industrial Change* (New York: Routledge, 1999), 125.

65 Robert Guy Matthews, "Steel Triples Its Net Income," *Wall Street Journal*, October 29, 2008, B2.

66 Peter Marsh, "Mittal Fatigue," *Financial Times*, October 31, 2008, 8.

67 James Hartigan, Sreenivas Kamma, and Phillip Perry, "Are Subsidies More Dangerous Than Dumping? Evidence from Wealth Effects for the Steel Industry," *Journal of Economic Integration* 9 (March 1994): 47, 58.

68 Peter Dungan, *Rock Solid: The Impact of the Mining and Primary Metals Industries on the Canadian Economy* (Toronto: Institute for Policy Analysis, 1997), 41-43.

69 Gene M. Grossman, "Strategic Export Promotion: A Critique," in *Strategic Trade Policy*, ed. Paul R. Krugman (Cambridge, MA: MIT Press), 56.

70 Klaus Stegemann, "Policy Rivalry Among Industrial States: What Can We Learn from Models of Strategic Trade Policy?" *International Organization* 41 (Winter 1989): 91-92.

71 Jose Guilherme de Heraclito Lima, *Restructuring the U.S. Steel Industry: Semi-Finished Steel Imports, International Integration, and U.S. Adaptation* (Boulder: Westview, 1991), 3.

72 William H. Branson and Alvin K. Klevorick, "Strategic Behaviour and Trade Policy," in *Strategic Trade Policy and the New International Economics*, ed. Paul R. Krugman (Cambridge, MA: MIT Press, 1986), 248; Willig, "Economic Effects of Antidumping Policy," 71.

73 Viner, *Dumping*, 139-41.

74 Alan O. Sykes, "Antidumping and Antitrust: What Problems Does Each Address?" in *Brookings Trade Forum 1998*, ed. Robert Z. Lawrence (Washington: Brookings Institution), 32-34.
75 Ethier, "Dumping," 493.
76 Cooper, "Comments on Alan O. Sykes," 44-45.
77 John R. Baldwin and Richard E. Caves, "International Competition and Industrial Performance," in *The Economics and Politics of International Trade, Vol. II: Freedom and Trade*, ed. Gary Cook (London: Routledge, 1998), 60-80.
78 Henryk Kierzkowski, "Recent Advances in International Trade Theory: A Selective Survey," *Oxford Review of Economic Policy* 3 (Spring 1987): 12.
79 Andrew Dick, "Does Import Protection Act as Export Promotion? Evidence from the United States," in *Structural Change, Industrial Location, and Competitiveness*, ed. Joanne E. Oxley and Bernard Yeung (Cheltenham: Elgar, 1998), 495-96.
80 *Predatory Pricing* (Paris: OECD, May 31, 1989), 6.
81 James Hartigan, "Predatory Dumping," *Canadian Journal of Economics* 29 (February 1996): 229-30.
82 Willig, "Economic Effects of Antidumping Policy," 65.
83 *Predatory Pricing*, 5.
84 Ibid. 19.
85 Ibid., 5.
86 Ibid., 5.
87 Giulio M. Gallarotti, "Toward a Business-Cycle Model of Tariffs," *International Organization* 39 (Winter 1985): 158.
88 William R. Thomson and Rafael Reuveny, "Tariffs and Trade Fluctuations: Does Protectionism Matter as Much as We Think?" *International Organization* 52 (Spring 1998): 437.
89 Ronald Rogowski, "Trade and the Variety of Democratic Institutions," *International Organization* 41 (Spring 1987): 207-8, 223.
90 Helen Milner, "Resisting the Protectionist Temptation: Industry and the Making of Trade Policy in France and the United States in the 1970s," *International Organization* 41 (Fall 1987): 641-42.
91 Susan Strange, "The Management of Surplus Capacity," *International Organization* 33 (Summer 1979): 316, 335.
92 Milner, "Resisting the Protectionist Temptation," 645.
93 J. David Richardson, "The New Political Economy of Trade Policy," in *Strategic Trade Policy*, ed. Paul R. Krugman (Cambridge, MA: MIT Press), 261.
94 Oona Hathaway, "Positive Feedback: The Impact of Trade Liberalization on Industry Demands for Protection," *International Organization* 52 (Summer 1998): 580-83.
95 Richardson, "The New Political Economy of Trade Policy," 261.
96 Jeffrey A. Frieden and Ronald Rogowski, "The Impact of the International Economy on National Policies: An Analytical Overview," in *Internationalization and Domestic Politics*, ed. Robert O. Keohane and Helen V. Milner (New York: Cambridge University Press, 1996), 38, 46.
97 James Alt, Jeffrey Frieden, Michael Gilligan, Dani Rodrik, and Ronald Rogowski, "The Political Economy of International Trade: Enduring Puzzles and an Agenda for Inquiry," *Comparative Political Studies* 29 (December 1996): 692.
98 Jorge Miranda, Raul A. Torres, and Mario Ruiz, "The International Use of Antidumping: 1987-1997," *Journal of World Trade* 32 (October 1998): 19.
99 William B.P. Robson, S. Dahlia Stein, and Rafael Fernandez de Castro, "What's the Fight About? An Overview of Trade Disputes in North America," in *Trading Punches: Trade Remedy Law and Disputes under NAFTA*, ed. Beatriz Leycegui, William B.P. Robson, and S. Dahlia Stein (Washington: National Planning Association, 1995), 11.
100 Richardson, "The New Political Economy of Trade Policy," 261.
101 Daniel M. Bernhofen, "Price Dumping in Intermediate Goods Markets," *Journal of International Economics* 39 (August 1995): 160.
102 Bela Belassa, "Tariff Reductions and Trade Among the Industrial Countries," *American Economic Review* 56 (June 1966): 469.

103   Michael O. Moore, "Waning Influence of Big Steel?" in *The Political Economy of American Trade Policy,* ed. Anne O. Krueger (Chicago: University of Chicago Press, 1996), 84.
104   Bernhofen, "Price Dumping in Intermediate Goods Markets," 160-61.
105   Faten Sabry, "An Analysis of the Decision to File, the Dumping Estimates, and the Outcome of Antidumping Petitions," *International Trade Journal* 14 (Summer 2000): 119-20.
106   Sykes, "Antidumping and Antitrust," 41.
107   Helen V. Milner, *Interests, Institutions, and Information: Domestic Politics and International Relations* (Princeton: Princeton University Press, 1997), 33-34.
108   Ibid., 62-63.
109   Ibid., 34-37.
110   Gallarotti, "Toward a Business-Cycle Model of Tariffs," 157-58.
111   Milner, *Interests, Institutions, and Information,* 17.
112   Timothy McKeown, "Firms and Tariff Regime Change: Explaining the Demand for Protection," *World Politics* 36 (January 1984): 215-33.
113   Peter J. Katzenstein, Robert O. Keohane, and Stephen D. Krasner, "*International Organization* and the Study of World Politics," *International Organization* 52 (Autumn 1998): 669.
114   Frieden and Rogowski, "The Impact of the International Economy on National Policies," 35, 43-44.
115   Goldstein, "The Political Economy of Trade," 166-67.
116   Barry Eichengreen, "Dental Hygiene and Nuclear War: How International Relations Looks from Economics," *International Organization* 52 (Autumn 1998): 995-96.
117   Hathaway, "Positive Feedback," 583.
118   Geoffrey Garrett and Peter Lange, "Internationalization, Institutions, and Political Change," in *Internationalization and Domestic Politics,* ed. Robert O. Keohane and Helen V. Milner (New York: Cambridge University Press, 1996), 66-67.

Chapter 4: Trading Steel

1   IISI, *World Steel in Figures 2008,* World Trade in Steel Products 1975 to 2007 (Brussels: 2008), 20.
2   Peter J.B. Steele, "The Propensity of Advanced Free World Economies to Import Steel," *Canada-United States Law Journal* 2 (1979): 25.
3   IISI, *World Steel in Figures 2008,* World Trade in Steel Products 1975 to 2007, 20.
4   Steele, "The Propensity of Advanced Free World Economies," 27.
5   IISI, *World Steel in Figures 2008,* The Major Importers and Exporters of Steel 2006, 18.
6   Jacques Brozzetti, "Prospects for Steel in Building Construction," in *The Steel Industry in the New Millennium Vol. I: Technology and the Market,* ed. Ruggero Ranieri and Jonathan Aylen (London: IOM Communications), 1998. For a brief assessment of the EU, see Jean-Louis Clipet, "Steel Market Analysis," prepared for the 133 Steel Committee, European Commission General Direction Enterprise and Industry, March 2, 2005, accessed on November 16, 2005, at http://www.fme.nl.
7   IISI, *World Steel in Figures 2008,* Apparent Steel Use 2001 to 2007, 12.
8   Ibid.
9   Alexander Gurov, "Steel in the Russian Federation" (Geneva: International Labour Organization, September 2000), accessed on February 16, 2006, at http://www.ilo.org.
10   William Hogan, "Prospects for the Steel Industry into the New Millennium," in *The Steel Industry in the New Millennium, Vol. I: Technology and the Market,* ed. Ruggero Ranieri and Jonathan Aylen (London: IOM Communications, 1998), 322.
11   Janet Plume, "Steel Imports Bounce Back," *Journal of Commerce,* June 21, 2004, 1.
12   Trade data accessed from IC on August 14, 2008, at http://www.ic.gc.ca.
13   OECD Directorate for Science, Technology, and Industry, "The Outlook for Steel," Background and Issues paper for OECD Special Meeting at High Level on Steel Issues (Paris: January 12-13, 2005), 4.
14   Jonathan R. Woetzel, "Remaking China's Steel Industry," *Asian Wall Street Journal,* December 27, 2001, accessed on July 26, 2005, at http://www.mckinsey.com.
15   IISI, *World Steel in Figures 2008,* Apparent Steel Use per Capita 2001 to 2007, 13.

16 IISI, *World Steel in Figures 2007,* Major Importers and Exporters of Steel 2005, 10; *World Steel in Figures 2008,* Major Importers and Exporters of Steel, 2006, 18.

17 "EU Warns China to Cool Steel Production," *Financial Post,* April 4, 2007, FP8.

18 Sandra Buchanan, "Europe Under Fire," *Metal Bulletin Monthly,* November 2007, 9.

19 Jason Bush and Sonal Rupani, "Russia's Steel Wheels Roll into America," *Business Week,* October 1, 2007, 44.

20 IISI, *World Steel in Figures 2008,* Major Importers and Exporters of Steel, 2006, 18; Peter Marsh, "Steel Set for Fall as China Increases Output," *Financial Times,* November 6, 2005, accessed on November 16, 2005, at http://us.ft.com.

21 IISI, *World Steel in Figures 2008,* World Crude Steel Production, 1950 to 2007, 6.

22 IISI, *World Steel in Figures, 2004* (Brussels: 2004), 3; *World Steel in Figures 2008,* Major Steel-Producing Countries, 2006 and 2007, 8.

23 Ibid.

24 International Iron and Steel Statistics Bureau, accessed on July 28, 2006, at http://issb.co.uk.

25 U.S. Department of Census import and export database, accessed on July 28, 2008, at http://www.census.gov.

26 Canada trade database, accessed on August 7, 2008, at http://ic.gc.ca. The aggregate industrial classification for steel imports includes iron.

27 IISI, *World Steel in Figures 2008,* World Steel Exports – Analysis by Product, 2002 to 2006, 19.

28 U.S. Census Bureau, International Trade Statistics. Accessed on August 8, 2008, at http://www.census.gov. The aggregate steel classification includes iron.

29 Paul R. Krugman, *Geography and Trade* (Leuven and Cambridge, MA: Leuven University Press and MIT Press, 1991), 4-6.

30 Donald R. Davis and David E. Weinstein, "Economic Geography and Regional Production Structure: An Empirical Investigation," *European Economic Review* 43 (February 1999): 381.

31 Krugman, *Geography and Trade,* 7.

32 Anthony D'Costa, *The Global Restructuring of the Steel Industry: Innovations, Institutions, and Industrial Change* (New York: Routledge, 1999), 12-21.

33 B. O'hUallachain, "The Restructuring of the U.S. Steel Industry: Changes in the Location of Production and Employment," *Environment and Planning A* 26 (September 1993): 1340.

34 Karsten Junius, *The Economic Geography of Production, Trade, and Development* (Tübingen: Mohr Siebeck, 1999), 138-39; a useful typology of industrial clusters appears in Peter Dicken, *Global Shift: Reshaping the Global Economic Map in the 21st Century,* 4th ed. (New York: Guilford, 2003), 22-23.

35 Dicken, *Global Shift,* 24.

36 Krugman, *Geography and Trade,* 49.

37 Ibid., 50-51.

38 A discussion of clusters, linkages, and economic theory appears in Paul R. Krugman, *Development, Geography, and Economic Theory* (Cambridge: MIT Press, 1995), 18-63.

39 John S. Hekman, "An Analysis of the Changing Location of Iron and Steel Production in the Twentieth Century," *American Economic Review* 68 (March 1978): 123-33.

40 O' hUallachain, "The Restructuring of the U.S. Steel Industry," 1355.

41 Metals and Minerals Processing Branch, *Primary Steel: Overview and Prospects* (Ottawa: Minister of Supply and Services, 1996), 18.

42 O' hUallachain, "The Restructuring of the U.S. Steel Industry," 1346-47.

43 Paul R. Krugman, "Where Is the 'New Economic Geography'?" in *The Oxford Handbook of Economic Geography,* ed. Gordon L. Clark, Maryann P. Feldman, and Meric S. Gertler (Oxford: Oxford University Press, 2000), 53.

44 Anthony Venables, "Equilibrium Locations of Vertically Linked Industries," *International Economic Review* 37 (May 1996): 341-59.

45 Richard Harris, "Globalization, Trade, and Income," *Canadian Journal of Economics* 26 (November 1993): 770.

46 Junius, *The Economic Geography of Production,* 141.

47 Paul R. Krugman, "Increasing Returns, Monopolistic Competition, and International Trade," *Journal of International Economics* 9 (November 1979): 479.

48 Davis and Weinstein, "Economic Geography and Regional Production Structure," 381-403.
49 Hans G. Mueller, "The Steel Industry," *Annals of the American Academy of Political and Social Science* 460 (March 1982): 77.
50 Ibid., 75.
51 Jose Guilherme de Heraclito Lima, *Restructuring the U.S. Steel Industry: Semi-Finished Steel Imports, International Integration, and U.S. Adaptation* (Boulder: Westview, 1991), 103-4.
52 Kent Jones, "Structural Adjustment in the United States Steel Industry," in *Industry on the Move: Causes and Consequences of International Relocation in the Manufacturing Industry*, ed. Gijsbert van Liemt (Geneva: International Labour Organization, 1992), 188.
53 C. Sherman Cheung, Itzhak Krinsky, and Bernadette Lynn, "The Canadian Steel Industry: Current State and Future Prospects," *Canadian Banker* 92 (April 1985): 9.
54 Metals and Minerals Processing Branch, *Primary Steel*, 2.
55 Myrna Pinkham, "As Good As It Gets," *Metal Bulletin Monthly*, January 2006, 24.
56 Robert Sherefkin, "Automakers to Get Steel Mill in South," *Automotive News*, May 3, 2004, 3.
57 Pinkham, "As Good As It Gets," 24; an informative interactive map of the facility can be found on Severcorr's website, http://www.severcorr.com.
58 Lindsay Chappell, "In Alabama, Hyundai Bets on Radical Korean Steel-Making Process," *Automotive News*, September 13, 2004, 1, 2.
59 OECD, "The Outlook for Steel," 5.
60 M.L. Dunn, *Recent Trends in the Seaborne Transportation of Iron and Steelmaking Raw Materials* (Brussels: IISI, 1992), 9-13.
61 IISI, *World Steel in Figures 2008*, Iron Ore 2006, 21.
62 OECD Business and Industry Policy Forum, Forum on Global Industrial Restructuring, Report: "Trends in Global Industrial Restructuring" (Paris: 19 February 2002).
63 OECD, "The Outlook for Steel," 5.
64 Anthony Poole, "Asia Steel Mills See Steady Freight on Top of Ore Price Rise," *Platt's International Coal Report*, June 19, 2006, 6.
65 William Kimberley, "The Steel Scene," *Automotive Design and Production*, February 2005, 18-19.
66 Kenneth Boyer, "Trucking, NAFTA, and the Cost of Distance," *Annals of the American Academy of Political and Social Science* 553 (September 1997): 59.
67 Jeffrey Rubin and Benjamin Tal, "Will Soaring Transport Costs Reverse Globalization?" *StrategEcon*, CIBC World Markets Inc., May 27, 2008, 4.
68 Ibid.
69 Donald Barnett, President, Economic Associates, at *New Steel* sixth annual roundtable on operating efficiencies, *New Steel*, August 1999.
70 Michael Storper, "Globalization, Localization, and Trade," in *The Oxford Handbook of Economic Geography*, ed. Gordon L. Clark, Maryann P. Feldman, and Meric S. Gertler (Oxford: Oxford University Press, 2000), 158.
71 Ibid., 159.
72 Junius, *The Economic Geography of Production*, 140.
73 Ibid.
74 Richard Baldwin, Rikard Foslid, Philippe Martin, Gianmarco Ottavian, and Frederic Robert-Nicout, *Economic Geography and Public Policy* (Princeton: Princeton University Press, 2003), 192-93.
75 Krugman, *Geography and Trade*, 50.
76 Storper, "Globalization, Localization, and Trade," 163.
77 Michael Storper, "Territories, Flows, and Hierarchies in the Global Economy," in *Reading Economic Geography*, ed. Tervor J. Barnes, Jamie Peck, Eric Sheppard, and Adam Ticknell (Malden: Blackwell, 2004), 282.
78 Ibid., 283.
79 Ibid., 284.
80 D'Costa, *The Global Restructuring of the Steel Industry*, 22-24.
81 Wim van Acker, "How to Deal with Steel," *Automotive Design and Production*, June 2005, 22.

82 Carlos Alberto Primo Braga, *Steel, Trade, and Development: A Comparative Advantage Analysis with Special Reference to the Case of Brazil*, Ph.D. dissertation, University of Illinois at Urbana-Champaign, 1984, 17.

83 Albert O. Hirschman, "The Political Economy of Import-Substitution Industrialization in Latin America," *Quarterly Journal of Economics* 82 (February 1968): 2-8.

84 Peter Dungan, *Rock Solid: The Impact of the Mining and Primary Metals Industries on the Canadian Economy* (Toronto: Institute for Policy Analysis, 1997), 144-50, 176-94.

85 Bernhard Fischer, Juan-Carlos Herken-Krauer, Matthias Lucke, and Peter Nunnenkamp, *Capital-Intensive Industries in Newly Industrializing Countries: The Case of the Brazilian Automobile and Steel Industries* (Tübingen: Mohr Siebeck, 1988), 164-65.

86 Braga, *Steel, Trade, and Development*, 14. A concise discussion and review of the relevant literature appears in Hajime Sato, "Total Factor Productivity vs. 'Realism': The Case of the South Korean Steel Industry," *Cambridge Journal of Economics* 29 (July 2005): 637-39.

87 Hiroshi Ohashi, "Learning by Doing, Export Subsidies, and Industry Growth: Japanese Steel in the 1950s and 1960s," *Journal of International Economics* 66 (July 2005): 303-4.

88 D'Costa, *The Global Restructuring of the Steel Industry*, 18-28.

89 Roger S. Ahlbrandt, Richard J. Fruehan, and Frank Giarratani, *The Renaissance of American Steel: Lessons for Managers in Competitive Industries* (New York: Oxford University Press, 1996), 12.

90 D'Costa, *The Global Restructuring of the Steel Industry*, 25.

91 Pei Sun, "Industrial Policy, Corporate Governance, and the Competitiveness of China's National Champions: The Case of Shanghai Baosteel Group," *Journal of Chinese Economic and Business Studies* 3 (May 2005): 174.

92 Moore, "Waning Influence of Big Steel?" 95.

93 Bernard Keeling, "Structural Change in the World Steel Industry: A North-South Perspective," in *Industry on the Move: Causes and Consequences of International Relocation in the Manufacturing Industry*, ed. Gijsbert van Liemt (Geneva: International Labour Organization, 1992), 171.

94 Ahlbrandt et al., *The Renaissance of American Steel*, 12.

95 Ibid.

96 Frances Stewart, "Recent Theories of International Trade: Some Implications for the South," in *Monopolistic Competition and International Trade*, ed. Henryk Kierzkowski (Oxford: Clarendon, 1984), 88.

97 Ohashi, "Learning by Doing, Export Subsidies, and Industry Growth," 302.

98 Ahlbrandt et al., *The Renaissance of American Steel*, 23.

99 Jang-Sup Chin, *The Economics of the Latecomers: Catching Up, Technology Transfer, and Institutions in Germany, Japan, and South Korea* (London: Routledge, 1996), 97.

100 Yoshitaka Suzuki, "The Rise and Decline of Steel Industries: A Business Historical Introduction," in *Changing Patterns of International Rivalry: Some Lessons from the Steel Industry*, ed. Etsuo Abe and Yoshitaka Suzuki (Tokyo: University of Tokyo Press, 1991), 11-12.

101 D'Costa, *The Global Restructuring of the Steel Industry*, 75, 77, 80.

102 For a management study of restructuring at Nippon Steel, Japan's largest integrated producer, see Ahlbrandt et al., *The Renaissance of American Steel*, 115-29.

103 "Japan Crude Steel Output Hits 34-Year Record," *Jiji Press English News Service*, April 21, 2008.

104 Anthony D'Costa, "State, Steel, and Strength: Structural Competitiveness and Development in South Korea," *Journal of Development Studies* 31 (October 1994): 53-59.

105 Kwang Soo Park and MoonJoong Tcha, "The Korean Steel Industry After the Economic Crisis: Challenges and Opportunities," in *The Korean Economy at the Crossroads*, ed. Moon-Joong Tcha and Chung-Sok Suh (London: RoutledgeCurzon, 2003), 200.

106 IISI, *World Steel in Figures 2008*, The Major Steel-Producing Countries 2006 to 2007, Apparent Steel Use 2001 to 2007, 8, 12; The Major Importers and Exporters of Steel, 2006, 18.

107 Chin, *The Economics of the Latecomers*, 102-7.

108 Suzuki, "The Rise and Decline of Steel Industries," 13; Peter Marsh, "Asian Steelmakers' Deal Fuels Concern," *Financial Times*, February 19, 2008, 17.

109 D'Costa, *The Global Restructuring of the Steel Industry*, 71-72.

110 Park and Tcha, "The Korean Steel Industry After the Economic Crisis," 202; IISI, *World Steel in Figures 2008,* Crude Steel Production by Process 2007, 9.
111 IISI, *World Steel in Figures 2008,* Top Steel-Producing Companies 2006 and 2007, 7.
112 Ian Zack, "Solid Steel Foundation," *Forbes,* July 22, 2002, 142.
113 Ibid.; "Posco Reports Record Quarterly Profit," *International Herald Tribune,* accessed on August 11, 2008, at www.iht.com.
114 Song Jung-A, "POSCO Looks for Chinese Partner," *Financial Times,* November 12, 2007, 20.
115 Steve Karpel, "Heavy Traffic," *Metal Bulletin Monthly,* June 2006, 32.
116 Song, "POSCO Looks for Chinese Partner," 20.
117 Paul Glader, "POSCO Looks Abroad to Grow; South Korean Steel Giant Confronts Industry Fears of Overcapacity," *Wall Street Journal,* October 31, 2005, B2.
118 Peter Marsh, "Arcelor in Talks to Work with POSCO," *Financial Times,* March 29, 2007, 26.
119 Park and Tcha, "The Korean Steel Industry After the Economic Crisis," 208-10.
120 "POSCO Reports Record Quarterly Profit, *International Herald Tribune,* July 11, 2008."
121 Vince Courteny, "Steely Resolve," *Ward's Auto World,* September 2005, 30.
122 D'Costa, *The Global Restructuring of the Steel Industry,* 86.
123 Fischer et al., *Capital-Intensive Industries in Newly Industrializing Countries,* 266.
124 Michael O. Moore, "Waning Influence of Big Steel?" in *The Political Economy of American Trade Policy,* ed. Anne O. Krueger (Chicago: University of Chicago Press, 1996), 95.
125 Edmund Amann and Frederick Nixson, "Globalisation and the Brazilian Steel Industry: 1988-1997," *Journal of Development Studies* 35 (August 1999): 80.
126 Ibid., 63.
127 Alfred Montero, "State Interests and the New Industrial Policy in Brazil: The Privatization of Steel, 1990-1994," *Journal of Interamerican Studies and World Affairs* 40 (Fall 1998): 27-63.
128 Philip Siekman, "Good Steel Made Cheaply," *Fortune* 147 (May 12, 2003): 144B.
129 Amann and Nixson, "Globalisation and the Brazilian Steel Industry," 78.
130 Ibid., 71.
131 Ibid., 83.
132 Siekman, "Good Steel Made Cheaply," 144B.
133 IISI, *World Steel in Figures 2008,* Iron Ore 2006, 21.
134 IISI, *World Steel in Figures 2006,* Scrap Estimated Consumption, Trade and Apparent Domestic Supply, 2004, 17; *World Steel in Figures 2008,* Scrap: Estimated Consumption, Trade and Apparent Domestic Supply, 2007, 23.
135 Amann and Nixson, "Globalisation and the Brazilian Steel Industry," 73, 78.
136 IISI, *World Steel in Figures 2008,* Major Importers and Exporters of Steel 2006, 18.
137 Siekman, "Good Steel Made Cheaply," 144B.
138 Diana Kinch, "Brazilian Demand Sparks Imports," *Metal Bulletin Monthly,* July-August 2007, 439.
139 IISI, *World Steel in Figures 2008,* Top Steel-Producing Companies 2006 and 2007, 7.
140 IISI, *World Steel in Figures 2005,* Top Steel-Producing Companies 2003 and 2004, 2; Siekman, "Good Steel Made Cheaply."
141 Claudio Mendonca, "Usiminas Announces US$14 bn 2008-12 Investment Package," *Business News Americas,* July 8, 2008, accessed on August 14, 2003, at http://www.bnamericas.com.
142 Ibid.; IISI, *World Steel in Figures 2008,* Top Steel-Producing Companies 2006 and 2007, 7.
143 U.S. Census Bureau, International Trade Statistics, accessed on August 14, 2008, at http://censtats.census.gov.
144 Ibid.
145 Stephen Cooney, *Current Issues in the Steel Industry* (New York: Novinka, 2003), 14.
146 Kent Jones, "Structural Adjustment in the United States Steel Industry," 183.
147 John Sheridan, "A Global Future?" *Industry Week* 248 (January 18, 1999): 125-26.
148 Greg Rushford, "Bush Steps into a Steel Trap," *Wall Street Journal,* March 6, 2002, A16.

149 Nelson D. Schwartz, "Bent but Unbowed," *Fortune*, July 22, 2002, 118.
150 "High Costs a Factor in Dofasco's Disappointing Third Quarter," *St. Catharines Standard*, November 1, 2005, D3; Reg Curren, "Dofasco Earnings Fall 71 Percent as Prices Decline, Costs Rise," *Ottawa Citizen*, February 4, 2006, D3.
151 "Brazil: Where the Slabs Come From," *Economist*, July 13, 2002, 57.
152 Mark Tran, "Corus Abandons Its 13bn Merger Plan," *The Guardian*, accessed on November 13, 2002, at www.guardian.co.uk.
153 Sun, "Industrial Policy, Corporate Governance," 188; Todd Benson, "China Fuels Brazil's Dream of Being a Steel Power," *New York Times*, May 21, 2004, W7.
154 Ibid; Andrea Welsh, "Companies Around the World Look to Brazil for Their Iron Ore," *Wall Street Journal*, April 7, 2004, 1.
155 "Brazil's CVRD, China's Baosteel to Build Steel Plant," *Wall Street Journal*, August 24, 2007.
156 Peter Marsh, "Corus Tunes into the Overtures from a Russian Duo," *Financial Times*, March 20, 2004, 3.
157 Welsh, "Companies around the World Look to Brazil," 1.
158 "Arcelor in Brazil," accessed on November 3, 2005, at http://www.arcelormittal.com.
159 Elizabeth Johnson, "Laggard Turns into a Leader," *Financial Times*, June 14, 2006, 5.
160 Cooney, *Current Issues in the Steel Industry*, 14.
161 "Brazil's CSN to Buy Bankrupt Heartland Steel," *Iron Age New Steel* 17 (August 2001): 8.
162 Johnson, "Laggard Turns into a Leader."
163 Diana Kinch, "Latin America's Costs Deter," *Metal Bulletin Monthly*, April 2006, 10.
164 Sheridan, "A Global Future?"
165 "Janik Mehta, Steel's Still-Growing Giant," *Industry Week*, January 18, 1999, 120-23.
166 Diana Kinch, "Latin America: Gerdau Springs Forward," *Metal Bulletin Monthly*, May 2006, 10.
167 Luiz Andre Rico Vicente, "Latin America Beckons," *Metal Bulletin Monthly*, April 2006, 57.
168 Ibid.
169 Paul E. Lydolph, *Geography of the U.S.S.R.* (New York: Wiley, 1970), 505-18.
170 Matthew J. Sagers, "The Iron and Steel Industry in Russia and the CIS in the Mid-1990s," *Post-Soviet Geography and Economics* 37 (1996): 197.
171 Ibid.
172 John Thornhill, "Survival of the Fattest: Subsidies, Tax Breaks, and Favouritism Keep Much of Russian Industry Afloat," *Financial Times*, October 19, 1999, 22.
173 William W. Lewis, "In Russia's Economy, It's Survival of the Weakest," *Wall Street Journal*, November 4, 1999, A30; Thornhill, "Survival of the Fattest."
174 John Helmer, "US Investors Desert Big Russian Steelmaker," *Journal of Commerce*, May 7, 1999, 4A.
175 "Looking Abroad: Russian Steel," *The Economist*, May 28, 2005, 78.
176 Charles Clover, "Russian Oligarchs Take the Bankruptcy Route to Expansion," *Financial Times*, July 25, 2000, 3.
177 Sandra Buchanan, "Evraz Reaches Critical Mass," *Metal Bulletin Monthly*, June 2003, 43-44.
178 Sandra Buchanan, "Second Tier Choices," *Metal Bulletin Monthly*, May 2007, 24.
179 Neil Buckley, Joanna Chung, and Peter Marsh, "Steppe Change – Russia's Smart New Business Breed Is Looking West for Deals," *Financial Times*, November 21, 2006, 13.
180 Sagers, "The Iron and Steel Industry in Russia," 201.
181 "Russian Pre-Export Steel Loan Gains Strong Bank Response," *International Trade Finance*, May 13, 1998, 3.
182 Helene Cooper, "Fortress Europe Persists Despite Approach of Euro," *Wall Street Journal*, November 18, 1998, A19; Greg Mastel, "Russia, Steel, and US Policy," *Journal of Commerce*, January 4, 1999, 4A.
183 Nancy Dunne, "Russia Agrees to Hefty Cuts in Steel Exports," *Financial Times*, February 23, 1999, 3.
184 IISI, *World Steel in Figures 2008*, Top Steel-Producing Companies, 2006 and 2007, 7.
185 Simon Pirani, "Russians Ponder Where to Invest Next," *Metal Bulletin Monthly*, April 2005, 41.

186 Sandra Buchanan, "Fighting Fit," *Metal Bulletin Monthly,* May 2006, 19.
187 Ibid.
188 Elena Anankina, "Rating Russian Steel Companies," *Metal Bulletin Monthly,* June 2005, 61.
189 Buchanan, "Fighting Fit," 20.
190 Yu Qiao, "Steel Companies Aim to Target More Countries," *China Daily,* August 2, 2005, 9.
191 Jeffrey Marshall, "Inside a Steel Deal," *Financial Executive,* December 2005, 29-30.
192 "Severstal Challenges Essar's Acquisition of Esmark," *Metal Center News,* June 2008, 83; "Severstal Emerges as Winner in Esmark Tug-of-War," *Metal Center News,* July 2008, 34.
193 Isabel Gorst and Peter Marsh, "Physicist Intent on Proving Steel Can Float," *Financial Times,* May 6, 2005, 32.
194 "Evraz Group Buys Claymont Steel," *Metal Producing and Processing,* January-February 2008, 46.
195 Bush and Rupani, "Russia's Steel Wheels," 44.
196 "Looking Abroad: Russian Steel," *The Economist,* May 28, 2005, 78.
197 "Russian Pundits Not Impressed with Putin's Attack on Steel Plant," *BBC Monitoring Former Soviet Union,* July 25, 2008.
198 "Mechel Bashing," *The Economist,* August 2, 2008, 65-66.
199 Yanrui Wu, "China's Metals Industry (I)," in *The Economics of the East Asia Steel Industries,* ed. Yanrui Wu (Aldershot: Ashgate, 1998), 121.
200 Sun, "Industrial Policy, Corporate Governance," 180.
201 Gary H. Jefferson, "China's Iron and Steel Industry," *Journal of Development Economics* 33 (October 1990): 348.
202 They are, respectively, Baosteel, Anshan-Benxi, Jiangsu Shagang, Tangshan, and Wuhan.
203 Pei Sun, "Is the State-Led Industrial Restructuring Effective in Transition China? Evidence from the Steel Sector," *Cambridge Journal of Economics* 31 (July 2007): 606.
204 IISI, *World Steel in Figures 2008,* Top Steel-Producing Companies 2006 and 2007, 7; The Major Steel-Producing Countries 2007, 8.
205 Sun, "Is the State-Led Industrial Restructuring Effective?" 608-10.
206 Paul Denlinger, "Baosteel Moves to Secure Brazilian Iron Ore Sources with JV," accessed on April 7, 2005, at http://www.china-ready.com/news/Feb2004.
207 Sun, "Industrial Policy, Corporate Governance," 188.
208 Patrick Barta and Paul Glader, "China's Steel Threat May Be Excess, Not Shortage," *Wall Street Journal,* December 30, 2004, A1.
209 OECD Directorate for Science, Technology, and Industry, "Recent Steel Market Developments," prepared for the OECD Special Meeting at High Level on Steel Issues, June 28, 2004 (Paris: June 28, 2004), 2.
210 "From Accelerator to Brake," *The Economist,* October 8, 2005, 81.
211 "Lovingly Touched by Mao," *The Economist,* February 2, 2002, 63.
212 Bartha and Glader, "China's Steel Threat May Be Excess," A1.
213 IISI, *World Steel in Figures 2006,* Major Importers and Exporters of Steel, 12; *World Steel in Figures 2008,* Major Importers and Exporters of Steel, 2006, 18.
214 http://steelonthenet.com/prices.html, accessed on July 28, 2006.
215 Anthony Poole, "Chinese Steel Demand Growth to Slow for Remainder of Decade," *Platt's International Coal Report,* March 27, 2006, 11.
216 Bartha and Glader, "China's Steel Threat May Be Excess," A1.
217 Ibid.
218 "From Accelerator to Brake," *The Economist,* October 8, 2005, 81, 82.
219 Paul Krugman, "A Model of Innovation, Technology Transfer, and World Distribution of Income," *Journal of Political Economy* 87 (April 1979): 253-66; James Kurth, "The Political Consequences of the Product Cycle," *International Organization* 33 (Winter 1979): 1-34.
220 Fischer et al., *Capital-Intensive Industries in Newly Industrializing Countries,* 3-4.
221 Robert R. Crandall, *The U.S. Steel Industry in Recurrent Crisis: Policy Options in a Competitive World* (Washington: Brookings Institution, 1981), 33-34.
222 Jeffrey A. Frieden and Ronald Rogowski, "The Impact of the International Economy on National Policies: An Analytical Overview," in *Internationalization and Domestic Politics,* ed.

Robert O. Keohane and Helen V. Milner (New York: Cambridge University Press, 1996), 38, 46.

223 Harris, "Globalization, Trade, and Income," 768; Crandall, *The U.S. Steel Industry,* 34-35; Fischer et al., *Capital-Intensive Industries in Newly Industrializing Countries,* 164-65.

224 Dungan, *Rock Solid,* 27.

225 Kenneth Warren, *World Steel: An Economic Geography* (New York: Crane, Russak, 1975), 267.

226 Donald Barnett, President, Economic Associates, at *New Steel* sixth annual roundtable on operating efficiencies, *New Steel,* August 1999.

227 Robert R. Miller, "The Changing Economics of Steel," *Finance and Development* 28 (June 1991): 39.

228 D'Costa, *The Global Restructuring of the Steel Industry,* 4.

229 Andrew Sharkey, "US Steel CEOs Confident but Concerned," *Industry Week,* March 2006, 15.

230 Lila J. Truett and Dale B. Truett, "The Korean Metals Industry and Economic Development," *Journal of Asian Economics* 3 (1997): 342; Miller, "The Changing Economics of Steel," 39.

231 Ibid.

232 A UBS comparative cost estimate for 2005 for a tonne of hot-rolled steel shows China's at $325. Even so, China ranks eighth of nineteen world producers, with costs greater than Taiwan's and just below Sweden's. Buchanan, "Fighting Fit," 20.

233 Tim Triplett, "China's Not Necessarily the Evil You Imagine," *Metal Center News,* March 2006, 6l; Sharkey, "US Steel CEOs Confident but Concerned," 15.

234 Sung-il Juhn, "Challenge of a Latecomer: The Case of the Korean Steel Industry with Specific Reference to POSCO," in *Changing Patterns of International Rivalry: Some Lessons from the Steel Industry,* ed. Etsuo Abe and Yoshitaka Suzuki (Tokyo: University of Tokyo Press, 1991), 276-77.

235 Fischer et al., *Capital-Intensive Industries in Newly Industrializing Countries,* 194-98, 206, 267; Braga, *Steel, Trade, and Development,* 320.

236 D'Costa, *The Global Restructuring of the Steel Industry,* 107.

237 Braga, *Steel, Trade, and Development,* 14.

238 D'Costa, *The Global Restructuring of the Steel Industry,* 127.

239 Sun, "Industrial Policy, Corporate Governance," 180-81.

240 Kinch, "Latin America's Costs Deter," 10.

241 Sun, "Industrial Policy," 187.

242 Robert Guy Matthews, "Melting Benefits: Retiree Costs Drive Big Change in Steel," *Wall Street Journal,* April 25, 2002, A1.

243 This explanation was popularized by Raymond Vernon in *Sovereignty at Bay* (New York: Basic, 1971).

244 Robert Gilpin, *Global Political Economy: Understanding the International Economic Order* (Princeton: Princeton University Press, 2001), 144.

245 Adrian Wood, *North-South Trade, Employment, and Inequality: Changing Fortunes in a Skill-Driven World* (Oxford: Clarendon, 1994), 45-46.

246 Gilpin, *Global Political Economy,* 140.

247 Wood, *North-South Trade,* 45.

248 Robert Gilpin, *The Political Economy of International Relations* (Princeton: Princeton University Press, 1987), 96.

249 Ibid.

250 Ibid., 234-35.

251 A brief typology of the two sets of purposes appears in Dicken, *Global Shift,* 275.

252 Gilpin, *Global Political Economy,* 130.

253 Robert Gilpin, *The Challenge of Global Capitalism: World Economy in the 21st Century* (Princeton: Princeton University Press, 2000), 165-66.

254 Ibid., 168.

255 Ethan Kapstein, "Winners and Losers in the Global Economy," *International Organization* 54 (Spring 2000): 359-84.

256 Wood, *North-South Trade,* 10.

257 Gilpin, *Global Political Economy*, 293.
258 Paul R. Krugman, "Increasing Returns, Imperfect Competition, and the Positive Theory of International Trade," in *Handbook of International Economics, Vol. 3*, ed. Gene M. Grossman and Kenneth Rogoff (Amsterdam: North-Holland), 1995.
259 Gilpin, *The Challenge of Global Capitalism*, 165-68.
260 A clear depiction of the levels of transnational vertical integration appears in Dicken, *Global Shift*, 245-50.

Chapter 5: Survival

1   OECD Directorate for Science, Technology, and Industry, "Recent Steel Market Developments," prepared for the OECD Special Meeting at High Level on Steel Issues, June 28, 2004. Paris: 2004, 3.
2   John D. Allen, "Dumping and the Canadian Steel Industry," *Canada–United States Law Journal* 2 (1979): 45.
3   Price information accessed on August 19, 2008, at http://steelonthenet.com.
4   Jim Forbes, "Merger Myopia," *Metal Bulletin Monthly*, November 2005, 16.
5   Michael Storper, "Globalization, Localization, and Trade," in *Oxford Handbook of Economic Geography*, ed. Gordon L. Clark, Maryann Feldman and Meric S. Gertler (Oxford: Oxford University Press, 2000), 158.
6   Anthony D'Costa, *The Global Restructuring of the Steel Industry: Innovation, Institutions, and Industrial Change* (New York: Routledge, 1999), 121, 123.
7   Federal Reserve Bank of Cleveland, "The Steel Industry, Economic Trends," April 2000, 15.
8   James Harrigan, "The Impact of the Asia Crisis on U.S. Industry: An Almost-Free Lunch?" *Federal Reserve Bank of New York Economic Policy Review* (September 2000): 77.
9   Nelson Schwartz, "Bent but Unbowed," *Fortune*, July 22, 2002, 118.
10 William Hogan, "1999 Annual Commodities: Iron and Steel," *Engineering and Mining Journal*, March 1999, 29-30; Stephen Cooney, *Current Issues in the Steel Industry* (New York: Novinka, 2003), 4. This study originally appeared as *Steel Industry and Trade Issues*, Report for Congress RL31007 (Washington: Congressional Research Service, October 10, 2002).
11 Hogan, "1999 Annual Commodities," 29.
12 Gary Clyde Hufbauer and Ben Goodrich, "Steel: Big Problems, Better Solutions," Policy Brief 01-9 (Washington: Institute for International Economics, July 2001), accessed on October 1, 2006, at http://www.iie.com/publications/pb/pb.cfm?ResearchID=77.
13 Federal Reserve Bank of Cleveland, "The Steel Industry, Economic Trends," 16.
14 Ibid., 15.
15 Federal Reserve Bank of Cleveland, "The Steel Industry, Economic Trends," March 2001, 15.
16 Andy Shinnan and Kristina Shimmons, "Primary Metal Industries," Statistics Canada, Manufacturing, Construction, and Energy Division, April 2000, 3-4, accessed on November 23, 2005, at http://www.statcan.ca.
17 Ibid.
18 IC, *Primary Steel in Canada: An Industry Snapshot* (Ottawa: Industry Canada, 2000), 19.
19 "Helter Smelters," *The Economist*, May 25, 2002, 62; Schwartz, "Bent but Unbowed," 118.
20 Thomas W. Gerdel, "Steel Industry's Ills Taking Toll in Canada," *Cleveland Plain Dealer*, February 19, 2004, accessed on December 1, 2005, at http://www.cleveland.com/indepth/steel.
21 OECD, "Conclusions of the High Level Meeting on Steel, 17-18 September 2001" (Paris: September 18, 2001).
22 Bruce Stokes, "Steel Thyself, America," *National Journal*, March 9, 2002, 717.
23 Gary Clyde Hufbauer and Ben Goodrich, "Steel Policy: The Good, the Bad, and the Ugly," Policy Brief PB03-1 (Washington: Institute for International Economics, January 2003), 14.
24 OECD Directorate for Science, Technology, and Industry, "Recent Steel Market Developments," prepared for the OECD Special Meeting at High Level on Steel Issues, June 28, 2004 (Paris: 2004), 3.
25 Ibid., 2.
26 IISI, *World Steel in Figures 2008*, The Major Importers and Exporters of Steel, 2006 (Brussels: 2008), 18.

27 IC, *Primary Steel in Canada*, 19.
28 IISI, *World Steel in Figures 2008*, World Crude Steel Production, 1950 to 2007, 6; Apparent Steel Use 2001 to 2007, 12.
29 Hufbauer and Goodrich, "Steel Policy," 14.
30 Ibid.
31 Hufbauer and Goodrich, "Steel: Big Problems, Better Solutions."
32 Hufbauer and Goodrich, "Steel Policy," 15.
33 For steel import figures, see "WTO Leading Exporters and Importers of Iron and Steel 2006," accessed on August 20, 2008, at http://wto.org/. For GDP figures, see http://www.worldbank.org/WBSITE/EXTERNAL/DATASTATISTICS/0, accessed on August 20, 2008.
34 "Steel Price Crisis Demands Reason, Industry Solution," *Automotive News*, March 15, 2004, 12.
35 Myra Pinkham, "North America: Imports Stir Things Up," *Metal Bulletin Monthly*, January 2006, 11.
36 Mark Tatge, "Rusty Steelmakers Take a Shine," *Forbes*, July 5, 2004, 150.
37 IISI, *World Steel in Figures 2005*, 12; *World Steel in Figures 2006*, 12.
38 "Bumper Profit Margins for Steel," *PR Newswire Europe*, February 24, 2006.
39 Stanley Reed and Adam Aston, "The Mergers Aren't Over Yet," *Business Week*, February 21, 2005, 6.
40 "Boom and Bust," *Canadian Business*, Summer 2005, 103.
41 Paul Glader, "Global Steelmakers Plot Consolidation to Limit Capacity," *Wall Street Journal*, July 20, 2005, A6.
42 Geoff Dyer, "Upturn in Investment Activity Helps Prices Rebound," *Financial Times*, June 14, 2006, 2; BBC News, July 20, 2005, broadcast at 8:26 GMT.
43 IC, *Primary Steel in Canada: An Industry Snapshot*, 20.
44 Thomas Watson, "Heavy Metal," *Canadian Business*, November 21-December 4, 2005, 35.
45 John Greenwood, "Bust Could Be Worst Since End of Second World War," *Financial Post*, July 7, 2006, FP1, 4.
46 "Major Chinese Steel Producers See Profits More Than Double in First Half," *China View*, accessed on September 15, 2007, at http//www.chinaview.cn.
47 IISI, *World Steel in Figures 2006*, Top Steel-Producing Companies 2004 and 2005, 2; *World Steel in Figures 2008*, Top Steel-Producing Companies 2006 and 2007, 7.
48 "Rot in the Vaults," *The Economist*, April 8, 2006, 74.
49 Myra Pinkham, "Trade Friction Heats Up," *Metal Bulletin Monthly*, June 2006, 32.
50 David Berman, "Are the Dividends Made of Steel?" *Financial Post*, August 14, 2005, FP7.
51 Paul Glader, "Profits for Merging Mittal, Arcelor Decline as Costs Rise," *Globe and Mail*, August 8, 2006, B9.
52 Matthew Boyle, "Mittal Steel President Calls for Chinese Steel Consolidation," *Platt's International Coal Report*, December 5, 2005, 8.
53 Kevin Foster, "Asia: Chinese Weight," *Metal Bulletin Monthly*, April 2006, 10.
54 Raghavendra Upadhyana and Gurdev Singh Virk, "Tata Steel Moves to Try to Prevent Takeover Bid," *Wall Street Journal*, July 6, 2006, A12.
55 Peter Marsh, "Ask the Expert: Steel Industry Consolidation," *Financial Times*, February 6, 2007, 1.
56 "EU Warns China to Cool Steel Production," *Financial Post*, April 4, 2007, FP8.
57 IISI, *World Steel in Figures 2008*, Top Steel-Producing Companies, 2006 and 2007, 8.
58 Ibid.
59 Forbes, "Merger Myopia," 16.
60 Paul Glader, "POSCO Looks Abroad to Grow; South Korean Steel Giant Confronts Industry Fears of Overcapacity," *Wall Street Journal*, October 31, 2005, B2.
61 Tom Stundz, "Steel Buyers Will See More Imports in 2009, but Growth Still Uncertain," *Purchasing*, November 13, 2008, 6.
62 Thomas Prusa, "An Overview of the Impact of U.S. Unfair Trade Laws," in *Trading Punches: Trade Remedy Law and Disputes Under NAFTA*, ed. Beatriz Leycegui, William B.P. Robson, and S. Dahlia Stein (Washington: National Planning Association, 1995), 184.

63   Canada Border Services Agency, historical listing of dumping and subsidy investigations pursuant to the Special Import Measures Act, 1984, accessed on October 1, 2006, at http://www.cbsa-asfc.gc.ca.

64   Patrick A. Messerlin, "Antidumping, and Safeguards," in *The WTO After Seattle*, ed. Jeffrey J. Schott (Washington: Institute for International Economics, 2000), 163; "Unfair Protection," *The Economist*, November 7, 1998, 75-76.

65   Barry Lacombe, "Steel: Unions, Industry, Governments, and Tribunals," *Canada-United States Law Journal* 30 (2004): 212-16.

66   Keith B. Anderson, "Agency Discretion or Statutory Direction: Decision Making at the U.S. International Trade Commission," *Journal of Law and Economics* 36 (October 1993): 915-35.

67   James R. Markusen, "Comment" [to Michael O. Moore, "The Waning Influence of Big Steel?"] in *The Political Economy of American Trade Policy*, ed. Anne O. Krueger (Chicago: University of Chicago Press, 1996), 129.

68   Alan O. Sykes, "Antidumping and Antitrust: What Problems Does Each Address?" in *Brookings Trade Forum, 1998*, ed. Robert Z. Lawrence (Washington: Brookings Institution, 1998), 15-16, 24.

69   Michael Hart, *A Trading Nation: Canadian Trade Policy from Colonialism to Globalization* (Vancouver: UBC Press, 2002), 183; Dan Ciuriak, "Anti-Dumping at 100 Years and Counting: A Canadian Perspective," *The World Economy* 28 (May 2005): 641-49.

70   Brink Lindsey, "The U.S. Antidumping Law: Rhetoric Versus Reality," *Journal of World Trade* 34 (January 2000): 2.

71   Messerlin, "China in the World Trade Organization," 163.

72   Christian A. Conrad, "Dumping and Anti-Dumping Measures from a Competition and Allocation Perspective," *Journal of World Trade* 36 (June 2002): 654-56.

73   Greg Mastel, "The U.S. Steel Industry and Antidumping Law," *Challenge* 42 (May-June 1999): 85.

74   Michael Webb, "Ambiguous Consequences of Anti-Dumping Laws," *Economic Inquiry* 30 (1992): 443.

75   Donald F. Barnett and Louis Schorsch, *Steel: Upheaval in a Basic Industry* (Cambridge: Ballinger, 1983), 50.

76   Daniel Schwanen, "When Push Comes to Shove: Quantifying the Continuing Use of Trade 'Remedy' Laws Between Canada and the United States," in *Trading Punches: Trade Remedy Law and Disputes under NAFTA*, ed. Beatriz Leycegni, William B.P. Robson, and S. Dahlia Stein (Washington: National Planning Association, 1995), 167.

77   Gunnar Niels, "What Is Antidumping Policy Really About?" *Journal of Economic Surveys* 14 (September 2000): 478.

78   Prusa, "An Overview of the Impact of U.S. Unfair Trade Laws," 185, 198.

79   The early study is David G. Tarr, "Cyclical Dumping: The Case of Steel Products," *Journal of International Economics* 9 (January 1979): 61.

80   Jorge Miranda, Raul A. Torres, and Mario Ruiz, "The International Use of Antidumping: 1987-1997," *Journal of World Trade* 32 (October 1998): 19.

81   Markusen, "Comment," 129-30.

82   Michael O. Moore, "Made in America? U.S. Steelmaking in the 1990s: An American Resurgence?" *International Executive* 38 (July-August 1996), 433; Jae W. Chung, "Effects of U.S. Trade Remedy Law Enforcement Under Uncertainty: The Case of Steel," *Southern Economic Journal* 65 (July 1998): 152.

83   Michael O. Moore, "Steel Protection in the 1980s: The Waning Influence of Big Steel?" Working Paper No. 4760 (Cambridge: National Bureau of Economic Research, June 1994), 53; for a contrary view that emphasizes the voluntary steel quotas in effect in the early 1980s, see William R. Cline, "U.S. Strategic Trade and Industrial Policy: The Experience of Textiles, Steel, and Automobiles," in *Strategic Trade Policy and the New International Economics*, ed. Paul R. Krugman (Cambridge: MIT Press, 1986), 224.

84   Judith Goldstein, "The Political Economy of Trade: Institutions of Protection," *American Political Science Review* 80 (March 1986): 179.

85   Mike Greissel, "Steel's Trade Cases: '92 vs. '98," *New Steel*, November 1998.

86 Nancy E. Kelly, "The Debate over Antidumping-Case Data," *New Steel*, August 1998.
87 Hufbauer and Goodrich, "Steel: Big Problems, Better Solutions."
88 Kelly, "The Debate over Antidumping-Case Data."
89 On procedures under the act, see Mark Dutz, "Economic Impact of Canadian Antidumping Law," in *Brookings Trade Forum 1998*, ed. Robert Z. Lawrence (Washington: Brookings Institution, 1998), 100-3.
90 Canada Border Services Agency, historical listing.
91 Department of Finance Canada, *Anti-Dumping Information Paper*, November 17, 2004, accessed on October 1, 2006, at http://www.fin.gc.ca/.
92 Department of Finance Canada, *A Comparison of the Antidumping Systems of Canada and the USA*, accessed on October 2, 2006, at http://sice.oas.org/.
93 Lacombe, "Steel," 214.
94 Ibid., 212-16.
95 Phaedon Nicolaedes, "The Competitive Effects of Dumping," *Journal of World Trade* 24 (October 1990): 126; Robert W. Staiger and Frank Wolak, "The Effect of Domestic Antidumping Law in the Presence of Foreign Monopoly," *Journal of International Economics* 32 (1992): 278.
96 J. David Richardson, "The New Political Economy of Trade Policy," in *Strategic Trade Policy and the New International Economics*, ed. Paul Krugman (Cambridge: MIT Press, 1986), 259.
97 U.S. GAO, *International Monetary Fund: Trade Policies of IMF Borrowers*, Report B-282825 GAO/NSIAD/GGD-99-1784 (Washington: USGPO, June 1999), 14-16.
98 Michael O. Moore, "The Waning Influence of Big Steel?" in *The Political Economy of American Trade Policy*, ed. Anne O. Krueger (Chicago: University of Chicago Press, 1996), 100-1.
99 The OECD study is cited in Niels, "What Is Antidumping Policy Really About?" 483; and in J. Michael Finger in his review of Gabrielle Marceau, *Antidumping and Anti-Trust Issues in Free Trade Areas* (Oxford: Oxford University Press, 1994), published in *Journal of Economic Literature* 34 (December 1996): 1069.
100 Tarr, "Cyclical Dumping," 59.
101 Robert D. Willig, "Economic Effects of Antidumping Policy," in *Brookings Trade Forum 1998*, ed. Robert Z. Lawrence (Washington: Brookings Institution, 1998), 63.
102 Staiger and Wolak, "The Effect of Domestic Antidumping Law," 276, 278.
103 Schwanen, "When Push Comes to Shove," 171, 173.
104 Nicolaedes, "The Competitive Effects of Dumping," 124.
105 Hyun Ja Shin, "Possible Instances of Predatory Pricing in Recent U.S. Antidumping Cases," in *Brookings Trade Forum 1998*, ed. Robert Z. Lawrence (Washington: Brookings Institution, 1998), 94.
106 Nicolaides, "The Competition Effects of Dumping," 122-24.
107 William B.P. Robson, S. Dahlia Stein, and Rafael Fernandez de Castro, "What's the Fight About? An Overview of Trade Disputes in North America," in *Trading Punches: Trade Remedy Law and Disputes Under NAFTA*, ed. Beatriz Leycegni, William B.P. Robson, and S. Dahlia Stein (Washington: National Planning Association, 1995), 22.
108 Niels, "What Is Antidumping Really About?" 475-76.
109 Susan Hutton and Michael Trebilcock, "An Empirical Study of the Application of Canadian Anti-Dumping Laws: A Search for Normative Rationales," *Journal of World Trade* 24 (June 1990): 128, 130.
110 Robson, Stein, and Fernandez de Castro, "What's the Fight About?" 12.
111 Niels, "What Is Antidumping Really About?" 477.
112 Ibid.
113 Willig, "Economic Effects of Antidumping Policy," 67.
114 Ibid., 68.
115 Corinne Krupp and Susan Skeath, "Evidence on Upstream and Downstream Impacts of Antidumping Cases," *North American Journal of Economics and Finance*, 13 (August 2002): 172-74.
116 Michael D. Galloway, Bruce Blonigen, and Joseph E. Flynn, "Welfare Costs of U.S. Antidumping and Countervailing Duty Laws," *Journal of International Economics* 19 (December 1999) 236-37.
117 Moore, "The Waning Influence of Big Steel?" 84.

118 Schwanen, "When Push Comes to Shove," 171.
119 Philip Swagel, *Union Behaviour, Industry Rents, and Optimal Policies* (Washington: International Monetary Fund, 1996), 22, 25.
120 Andrew R. Dick, "Does Import Protection Act as Export Promotion? Evidence from the United States," in *Structural Change, Industrial Location, and Competitiveness,* ed. Joanne E. Oxley and Bernard Yeung, (Cheltenham: Elgar, 1998), 507-8.
121 Krupp and Skeath, "Evidence on the Upstream and Downstream Impacts of Antidumping Cases," 165.
122 Faten Sabry, "An Analysis of the Decision to File, the Dumping Estimates, and the Outcome of Antidumping Petitions," *International Trade Journal* 14 (Summer 2000): 119.
123 Kenneth H. Kelly and Morris E. Morkre, "Do Unfairly Traded Imports Injure Domestic Industries?" *Review of International Economics* 6 (May 1998): 321, 323, 326.
124 Michel Glais, "Steel Industry," in *European Policies on Competition, Trade, and Industry: Conflict and Complementarities,* ed. Pierre Buiges, Alexis Jacquemin, and Andre Sapir (Aldershot: Elgar, 1995), 250-51.
125 Webb, "Ambiguous Consequences of Anti-Dumping Laws," 444.
126 Poonam Gupta, "Why Do Firms Pay Antidumping Duty?" Working Paper (Washington: International Monetary Fund, 1999), 4, 5.
127 Hylke Vandenbussche and Xavier Wauthy, "Inflicting Injury Through Product Quality: How European Antidumping Policy Disadvantages European Producers," *European Journal of Political Economy* 17 (March 2001): 104.
128 Bruce A. Blonigen, "Tariff-Jumping Antidumping Duties," Working Paper 7776 (Cambridge, MA: National Bureau of Economic Research, July 2000), 27.
129 Moore, "The Waning Influence of Big Steel?" 87.
130 Richard G. Harris, "Trade and Industrial Policy for a 'Declining' Industry: The Case of the U.S. Steel Industry," in *Empirical Studies of Strategic Trade Policy,* ed. Paul Krugman and Alasdair Smith (Chicago: University of Chicago Press, 1994), 146.
131 Krupp and Skeath, "Evidence on the Upstream and Downstream Impacts," 165-66.
132 "Standing Up to Steel," *The Economist,* June 26, 1999, 31.
133 Ibid.
134 Julie Hirschfeld, "Bush's Trade Agenda Starts Out Between Big Steel and a Hard Place," *Congressional Quarterly Weekly* 59 (February 24, 2001), 431.
135 Mark Murray, "Trying to Steel Labour's Heart?" *National Journal* 33 (June 30, 2001): 2100.
136 Benjamin H. Liebman, "Safeguards, China, and the Price of Steel," *Review of World Economics* 142 (July 2006): 354-73.
137 "Steel Industry Mergers Could Lead to Stabilization," http://www.platts.com, accessed July 21, 2006.
138 ArcelorMittal Annual Report 2007, 3-8.
139 Ibid., 6.
140 Directorate for Science, Technology, and Industry, "The Outlook for Steel," Background and Issues Paper for OECD Special Meeting at High Level on Steel Issues, Paris: OECD, January 12-13, 2005, 7.
141 "Steel Industry Mergers Could Lead to Stabilization," accessed on July 21, 2006, at http://www.platts.com.
142 Stanley Reed and Adam Aston, "The Mergers Aren't Over Yet," *Business Week,* February 21, 2005, 6.
143 Jonathan Aylen, "Trends in the International Steel Market," in *The Steel Industry in the New Millennium, Vol. I: Technology and the Market,* ed. Ruggero Ranieri and Jonathan Aylen (London: IOM Communications, 1998), 210-11.
144 Peter Warrian and Celine Mulhern, "Knowledge and Innovation in the Interface Between the Steel and Automotive Industries: The Case of Dofasco," *Regional Studies* 39 (April 2005): 168.
145 Alexi Mordashov, "Why My Money Is on a Global Expansion in Steel," *Financial Times,* June 16, 2006, 15.
146 "Iron Ore and Metallurgical Coal to Prove Resilient to Latest Downturn in Steel Prices," *Platt's International Coal Report,* July 4, 2005, 1.

147 Boyle, "Mittal Steel President Calls for Chinese Steel Consolidation," 8.

148 Paul Magnusson, "Grand Old Protectionists," *Business Week,* March 26, 2001, 14.

149 Peter Marsh, "Crisis Forces ArcelorMittal to Slash Output," *Financial Times,* November 6, 2008.

150 Peter Marsh, "Showing a New Spark," *Financial Times,* June 14, 2006, 1.

151 Radikh Kamath, "Arcelor Deal Could Transform Global Steel Industry," *Business Line,* June 27, 2006, 1.

152 "Steel Industry Consolidation Behind Jump in Pipeline Costs," *Natural Gas Week,* November 24, 2004, 1.

153 Adam Ashton and Michael Arndt, "A New Goliath in Big," *Business Week,* November 8, 2004, 47.

154 "Pick and Mix," *Metal Bulletin Monthly,* November 2005, 9.

155 Kent Jones, "Structural Adjustment in the United States Steel Industry," in *Industry on the Move: Causes and Consequences of International Relocation in the Manufacturing Industry,* ed. Gijsbert van Liemt (Geneva: International Labour Organization, 1992), 197-98.

156 Aylen, "Trends in the International Steel Market," 211.

157 Sandra Buchanan, "Fighting Fit," *Metal Bulletin Monthly,* May 2006, 19.

158 Forbes, "Merger Myopia," 16.

159 "Forging a New Shape," *The Economist,* December 10, 2005, 67.

160 Patrick Barta, "China Relents, Accepts 19 Percent Rise in Iron-Ore Price," *Wall Street Journal,* June 21, 2006, C6.

161 http://www.steelonthenet.com, accessed September 22, 2006.

162 *PR Newswire Europe,* February 24, 2006.

163 "Steel Industry Mergers Could Lead to Stabilization," accessed on August 21, 2006, at http:///www.platts.com.

164 Alex Fak, "Peaking Prices Whip Companies Toward Vertical Integration," *Financial Times,* April 13, 2005, 2.

165 Boyd Erman, "New Dofasco Bid Expected," *Financial Post,* November 24, 2005, FP1.

166 "Industry Association Opposes Foreign Acquisition of Baosteel," *SinoCast China Business Daily News,* May 1, 2006, 1; Paul Glader, "Global Steelmakers Plot Consolidation to Limit Capacity," *Wall Street Journal* July 20, 2005, A6.

167 Humeyra Pamuk, "As Prices Wallow, Steel Industry Could Consolidate," *International Herald Tribune,* December 10, 2008.

168 Robert Sherefkin, "GM: Big Steel Must Go Global," *Automotive News,* November 18, 2002, 1-2.

169 Paul Glader, "Politics and Economics: Steel-Sector Consolidation Worries Customers," *Wall Street Journal,* March 23, 2006, A8.

170 Wim Oude Weernink, "New Rules, Rising Costs Affect Material Choices," *Automotive News,* April 28, 2008, 16B.

171 Adam Ashton and Michael Arndt, "A New Goliath in Big."

172 Harry Stoffer, "Study: Steel Will Hurt Suppliers," *Automotive News,* February 21, 2005, 4.

173 Paul Glader, "Politics and Economics," A8.

174 Robert Sherefkin, "GM: Big Steel Must Go Global," *Automotive News,* November 18, 2002, 1-2.

175 Glader, "Politics and Economics," A8.

176 Ibid.

177 William Kimberley, "The Steel Scene," *Automotive Design and Production,* February 2005, 18-19.

178 Matthew Boyle, "McArthur Executive Says Steel Consolidation to Change Australian Coal Industry," *Platt's International Coal Report,* April 10, 2006, 1.

179 Raghavendra Upadhyana and Gurdev Singh Virk, "Tata Steel Moves to Try to Prevent Takeover Bid," *Wall Street Journal,* July 6, 2006, A12.

180 "Three Japan Steelmakers Unveil Joint Takeover-Defence Plan," *Jiji Press English News Service,* March 29, 2006, 1.

181 Mariko Sanchanta, "Defences Rest on Shaky Foundations in Japan," *Financial Times,* April 19, 2006, 23.

182 "Still Room for Realignment of Japanese Steel Industry: JFE's Bada," *Jiji Press English News Service*, June 13, 2006, 1.

183 Pei Sun, "Industrial Policy, Corporate Governance, and the Competitiveness of China's National Champions: The Case of Shanghai Baosteel Group," *Journal of Chinese Economic and Business Studies* 3 (May 2005): 176.

184 P.F. Marcus, "Ironies in Steel," in *The Steel Industry in the New Millennium, Vol. I: Technology and the Market*, ed. Ruggero Ranieri and Jonathan Aylen (London: IOM Communications, 1998), 16.

185 Nam-Hoon Kang and Kentaro Sakai, *New Patterns of Industrial Globalization: Cross-Border Mergers and Acquisitions and Strategic Alliances* (Paris: OECD, 2001), 91.

186 IC, *Primary Steel in Canada*, 20.

187 Production figures accessed on August 22, 2008, at http://oica.net.

188 Peter Dicken, *Global Shift: Reshaping the Global Economic Map in the 21st Century*, 4th ed. (New York: Guilford, 2003), 357-61.

189 Ibid., 362.

190 "Driving Change," *The Economist*, September 4, 2004, 15.

191 Myra Pinkham, "Automotive Outlook: Big Three See the Light," *Metal Center News*, April 2006, 16.

192 Lindsay Chappell, "MPG Angst Holds Promise for Japanese Suspension Supplier," *Automotive News*, June 5, 2006, 53.

193 Weernink, "New Rules, Rising Costs," 16B.

194 Wim van Acker, "How to Deal with Steel," *Automotive Design and Production*, June 2005, 22; Kyung-Hee Jung, "Task Shift of Automakers to Steel Suppliers in the Value Chain of Automotive Sheets," *International Journal of Automotive Technology and Management* 5 (2005): 217.

195 Mark Johnson, "New Steel Has Strength, Repairability," *Automotive Body Repair News*, January 2006, 8.

196 Van Acker, "How to Deal with Steel," 22.

197 James Mackintosh, "Drive Goes on to Cut Weight," *Financial Times*, November 6, 2002, 4.

198 Gary S. Vasilash, "Steel Strikes Back," *Automotive Design and Production*, March 2006, 74.

199 Kang and Sakai, *New Patterns of Industrial Globalization*, 90.

200 M. Kipping, "Cooperation Between Steel Producers and Steel Users: A Major Determinant of National Competitive Advantage," in *The Steel Industry in the New Millennium Vol. I: Technology and the Market*, ed. Ruggero Rainieri and Jonathan Aylen (London: IOM Communications, 1998), 228.

201 Ibid., 218-19.

202 Mariko Sanchanta, "Land of the Rising Sun," *Financial Times*, June 14, 2006, 4.

203 Jung, "Task Shift of Automakers," 217.

204 Harry Stoffer, "Study: Still Will Hurt Suppliers," *Automotive News*, February 21, 2005, 4.

205 Dicken, *Global Shift*, 368.

206 Ibid., 355.

207 Ibid., 377-78.

208 Kipping, "Cooperation Between Steel Producers and Steel Users," 221-22.

209 Jung, "Task Shift of Automakers," 217-18.

210 Joseph Ogando, "Strong Bodies," *Design News*, April 5, 2004, 32.

211 Vasilash, "Steel Strikes Back," 74.

212 Lindsay Chappell, "MPG Angst Holds Promise for Japanese Suspension Supplier," *Automotive News*, June 5, 2006, 53.

213 Ogando, "Strong Bodies," 32.

214 Kang and Sakai, *New Patterns of Industrial Globalization*, 13, 28.

215 Ibid., 42.

216 Ibid., 27, 109.

217 Johnson, "New Steel Has Strength," 8.

218 Ken Hijino and Peter Marsh, "Sumitomo and Corus Forge Steel Link," *Financial Times*, January 11, 2002, 28.

219 Christian Koehl, "Auto Ventures Push Steel Advances," *Metal Bulletin Monthly*, December 2005, 30.

220 "Automotive," *Metal Bulletin Monthly,* March 2005, 46.
221 William Dielun, "Steel Group Opening Up?" *Ward's Auto World,* September 2005, 32; Jung, "Task Shift of Automakers," 230.
222 Accessed on August 22, 2006, at http://www.arcelor.com/subsite/2004AnnualResults.
223 Koehl, "Auto Ventures Push Steel Advances," 30.
224 "Steel JV Will Serve Japanese Automakers," *Automotive News,* June 27, 2005, 24.
225 Sandra Buchanan, "Made to Measure," *Metal Bulletin Monthly,* February 2006, 42.
226 Warrian and Mulhern, "Knowledge and Innovation," 166-67.
227 Arcelor, annual report, 2004.
228 Jung, "Task Shift of Automakers," 229.
229 A. Moreau, "Intermaterial Competition Beyond Technological and Economic Factors," in *The Steel Industry in the New Millennium, Vol. I: Technology and the Market,* ed. Ruggero Ranieri and Jonathan Aylen (London: IOM Communications, 1998), 268-69.
230 Jung, "Task Shift of Automakers," 229.
231 Koehl, "Auto Ventures Push Steel Advances," 30.
232 Bradford Werale, "Steelmaker Looks to Eastern Europe," *Automotive News,* July 5, 2004, 27.
233 Dofasco Incorporated, annual report 2004, accessed on August 24, 2006, at http://www.dofasco.ca.
234 Mariko Sanchanta, "Land of the Rising Sun," *Financial Times,* June 14, 2006, 4; "Steel JV Will Serve Japanese Automakers," *Automotive News,* June 27, 2005, 24.
235 For an excellent if somewhat technical discussion, see B. Curtis Eaton, "The Elementary Economics of Social Dilemmas," *Canadian Journal of Economics* 37 (November 2004): 805-29.
236 Press release, U.S. Department of Justice, February 20, 2007.
237 Robert Guy Matthews, "Mittal Sells Mill in U.S. as Part of Arcelor Deal," *Wall Street Journal,* August 3, 2007, A6.

Chapter 6: Steel in a Global Perspective

1 Donald R. Davis and David E. Weinstein, "Economic Geography and Regional Production Structure: An Empirical Investigation," *European Economic Review* 43 (February 1999): 381.
2 Ibid., 402-3.
3 John F. Helliwell, *Globalization and Well-Being* (Vancouver: UBC Press, 2002), 30-31.
4 Ibid., 32.
5 Herman M. Schwartz, *States Versus Markets: History, Geography, and the Development of the International Political Economy* (New York: St. Martin's, 1994), 62, 63.
6 Ibid., 63.
7 Albert G. Hu, Gary H. Jefferson, and Qian Jinchang, "R and D and Technology Transfer: Firm-Level Evidence from Chinese Industry," *Review of Economics and Statistics* 87 (November 2005): 785.
8 "Chinese Trade a Money Loser for Some," *Manufacturing Engineering* 137 (October 2006): 30.
9 IISI, *World Steel in Figures 2008,* Top Steel-Producing Companies, 2006 and 2007 (Brussels: 2008), 7.
10 Lintong Feng, "China's Steel Industry: Its Rapid Expansion and Influence on the International Steel Industry," *Resources Policy* 20 (December 1994): 222-23.
11 Schwartz, *States Versus Markets,* 106.
12 "Essar Steel Consolidates Presence in N. America," *Businessline,* May 3, 2008.
13 Jeffrey Rubin and Benjamin Tal, "Will Soaring Transport Costs Reverse Globalization?" *StrategEcon,* CIBC World Markets Inc., May 27, 2008, 4.
14 Ibid.
15 Ibid.
16 Ibid., 5.
17 Ibid., 6.
18 Ibid.
19 Nicolas Van Praet, "Shipping Costs Steer Automakers from China," *Financial Post,* May 28, 2008, FP3.
20 IISI, *World Steel in Figures 2008,* Crude Steel Production by Process, 2007, 9.

21 Larry Rohter, "Shipping Costs Start to Crimp Globalization," *International Herald Tribune,* August 2, 2008, accessed on August 2, 2008, at http://www.iht.com.

22 Ibid.

23 Steven R. Weisman, "U.S. Fears Overseas Funds Could 'Buy Up America,'" *New York Times,* August 20, 2007.

24 Vladimir Socor, "Russia Cuts Supplies to Czech Republic Without Explanation," *Eurasia Daily Monitor,* July 15, 2008.

25 Ibid.; "Keep Your T-Bonds, We'll Take the Bank," *The Economist,* July 28, 2007, 75-76.

26 Neil Buckley, Joanna Chung, and Peter Marsh, "Steppe Change – Russia's Smart New Business Breed Is Looking West for Deals," *Financial Times,* November 21, 2006, 13.

27 Shawn McCarthy, "How Ottawa Let Steel Industry Slip Away," *Globe and Mail,* August 28, 2007, A1.

28 Paul Glader and Guy Chazan, "Russian Foundries Court the West," *Wall Street Journal,* November 24, 2006, C1; on close ties of the former KGB to Russian business, see "The Making of a Neo-KGB State," *The Economist,* August 25, 2007, 25-28; for a warning about the potential risks of dealing with Kremlin-favoured oligarchs, see "White Knight on a Trojan Horse," *Business Europe,* May 16-May 31 2006, 3.

29 "Russian Pundits Not Impressed with Putin's Attack on Steel Plant," *BBC Monitoring Former Soviet Union,* July 25, 2008.

30 Peter Koven, "'Sexy' Ipsco Courted by Russian Suitor," *Financial Post,* April 13, 2007, FP5.

31 "Oregon Steel Sale Closes," *Portland Business Journal,* January 23, 2007.

32 Stephanie Kirchgaessner, "Tougher Scrutiny of Foreign Takeovers in US," *Financial Times,* July 27, 2007, 5; Niels C. Sorrells and Andrew Peaple, "Politics and Economics: Germany Seeks EU Curbs on Some Foreign Takeovers," *Wall Street Journal,* July 19, 2007, A8.

33 The panel's report is available online at http://www.competitionreview.ca.

34 Weisman, "U.S. Fears Overseas Funds."

35 That can be seen in full measure in "Compete to Win," the report of the Competition Review Panel. Available at http://www.competitionreview.ca.

Bibliography

Ahlbrandt, Roger S., Richard J. Fruehan, and Frank Giarratani. *The Renaissance of American Steel: Lessons for Managers in Competitive Industries*. New York: Oxford University Press, 1996.

Allen, John D. "Dumping and the Canadian Steel Industry." *Canada–United States Law Journal* 2 (1970): 39-51.

Alt, James, Jeffrey Frieden, Michael Gilligan, Dani Rodrik, and Ronald Rogowski. "The Political Economy of International Trade: Enduring Puzzles and an Agenda for Inquiry." *Comparative Political Studies* 29 (December 1996): 689-717.

Amann, Edmund, and Frederick Nixson. "Globalisation and the Brazilian Steel Industry: 1988-1997." *Journal of Development Studies* 35 (August 1999): 59-88.

Anderson, Keith B. "Agency Discretion or Statutory Direction: Decision Making at the U.S. International Trade Commission." *Journal of Law and Economics* 36 (October 1993): 915-35.

Astier, Jacques E. "Evolution of the World Iron Ore Market." *Minerals and Energy* 16 (December 2001): 23-30.

Aylen, Jonathan. "Trends in the International Steel Market." In *The Steel Industry in the New Millennium, Vol. I: Technology and the Market*, ed. Ruggero Ranieri and Jonathan Aylen, 197-213. London: IOM Communications, 1998.

Baldwin, John R., and Richard E. Caves. "International Competition and Industrial Performance." In *The Economics and Politics of International Trade, Vol. II: Freedom and Trade*, ed. Gary Cook, 57-84. London: Routledge, 1998.

Baldwin, Richard, Rikard Foslid, Philippe Martin, Gianmarco Ottavian, and Frederic Robert-Nicout. *Economic Geography and Public Policy*. Princeton: Princeton University Press, 2003.

Barnett, Donald F., and Robert W. Crandall. "Steel: Decline and Renewal." In *Industry Studies*, 2nd ed., ed. Larry L. Deutsch, 124-43. Armonk, NY: M.E. Sharpe, 1998.

–. *Up from the Ashes: The Rise of the Steel Minimill in the United States*. Washington: Brookings Institution, 1986.

Barnett, Donald F., and Louis Schorsch. *Steel: Upheaval in a Basic Industry*. Cambridge, MA: Ballinger, 1983.

Baumann, H.G. "The Relative Competitiveness of the Canadian and U.S. Steel Industries, 1955-1970." *Economia Internazionale* 27 (February 1974): 141-56.

Belassa, Bela. "Tariff Reductions and Trade Among the Industrial Countries." *American Economic Review* 56 (June 1966): 466-73.

Bernhofen, Daniel M. "Price Dumping in Intermediate Goods Markets." *Journal of International Economics* 39 (August 1995): 159-73.

Bilous, Jarret, and Kam Hon. *North American Steel Industry Study*. Toronto: Dominion Bond Rating Service, March 2004.

Blonigen, Bruce A. "Tariff-Jumping Antidumping Duties." Working Paper 7776. Cambridge, MA: National Bureau of Economic Research, July 2000.

Bouquet, Tim, and Byron Ousey. *Cold Steel: Britain's Richest Man and the Multi-Billion-Dollar Battle for a Global Empire.* Boston: Little, Brown, 2008.

Boyer, Kenneth. "Trucking, NAFTA, and the Cost of Distance." *Annals of the American Academy of Political and Social Science* 553 (September 1997): 55-65.

Braga, Carlos Alberto Primo. *Steel, Trade, and Development: A Comparative Advantage Analysis with Special Reference to the Case of Brazil.* Ph.D. diss., University of Illinois at Urbana-Champaign, 1984.

Brander, James A. "Rationales for Strategic Trade and Industrial Policy." In *Strategic Trade Policy and the New International Economics,* ed. Paul R. Krugman, 23-46. Cambridge, MA: MIT Press, 1986.

Brander, James A., and Paul R. Krugman. "A 'Reciprocal Dumping' Model of International Trade." Discussion Paper #513. Kingston: Queen's University, Institute for Economic Research (n.d.).

Brannlund, Rudnar, and Karl-Gustaf Onfgren. "Cyclical Dumping and Correlated Business Cycles in Imperfect Markets: Empirical Applications to the Canadian Pulp and Paper Industry." *Applied Economics* 27 (1995) 1081-91.

Branson, William H., and Alvin K. Klevorick. "Strategic Behaviour and Trade Policy." In *Strategic Trade Policy and the New International Economics,* ed. Paul R. Krugman, 241-55. Cambridge, MA: MIT Press, 1986.

Brozzetti, Jacques. "Prospects for Steel in Building Construction," in *The Steel Industry in the New Millennium, Vol. I: Technology and the Market,* ed. Ruggero Ranieri and Jonathan Aylen, 275-87. London: IOM Communications, 1998.

Caves, Richard E., and Michael E. Porter. "Barriers to Exit." In *Essays on Industrial Organization in Honor of Joe S. Bain,* ed. Robert T. Masson and P. David Qualls, 31-69. Cambridge, MA: Ballinger, 1976.

Cheung, C. Sherman, Itzhak Krinsky, and Bernadette Lynn. "The Canadian Steel Industry: Current State and Future Prospects." *Canadian Banker* 92 (April 1985): 7-13.

Chin, Jang-Sup. *The Economics of the Latecomers: Catching Up, Technology Transfer, and Institutions in Germany, Japan, and South Korea.* London: Routledge, 1996.

Chung, Jae W. "Effects of U.S. Trade Remedy Law Enforcement Under Uncertainty: The Case of Steel." *Southern Economic Journal* 65 (July 1998): 151-59.

Ciuriak, Dan. "Anti-Dumping at 100 Years and Counting: A Canadian Perspective." *The World Economy* 28 (May 2005): 641-49.

Clancy, Peter. *Micropolitics and Canadian Business: Paper, Steel, and the Airlines.* Peterborough: Broadview, 2004.

Clarida, Richard. "Entry, Dumping, and Shakeout." *American Economic Review* 83 (March 1993): 180-202.

Cline, William R. "U.S. Strategic Trade and Industrial Policy: The Experience of Textiles, Steel, and Automobiles." In *Strategic Trade Policy and the New International Economics,* ed. Paul R. Krugman, 211-39. Cambridge, MA: MIT Press, 1996.

Conrad, Christian A. "Dumping and Anti-Dumping Measures from a Competition and Allocation Perspective." *Journal of World Trade* 36 (June 2002): 563-75.

Cooney, Stephen, *Current Issues in the Steel Industry.* New York: Novinka, 2003.

Cooper, Richard. "Comments on Alan O. Sykes, 'Antidumping and Antitrust: What Problems Does Each Address?'" In *Brookings Trade Forum 1998,* ed. Robert Z. Lawrence, 44-53. Washington: Brookings Institution, 1998.

Crandall, Robert. *The U.S. Steel Industry in Recurrent Crisis: Policy Options in a Competitive World.* Washington: Brookings Institution, 1981.

Dack, W.L. "Canada's Steel Industry Expands in a Big Way." *Canadian Geographical Journal* 91 (October 1975): 32-41.

Davies, Stephen W., and Anthony J. McGuinness. "Dumping at Less Than Marginal Cost." *Journal of International Economics* 21 (January 1982): 169-82.

Davis, Donald R., and David E. Weinstein. "Economic Geography and Regional Production Structure: An Empirical Investigation." *European Economic Review* 43 (February 1999): 379-407.

D'Costa, Anthony. *The Global Restructuring of the Steel Industry: Innovations, Institutions, and Industrial Change*. New York: Routledge, 1999.

–. "State, Steel, and Strength: Structural Competitiveness and Development in South Korea." *Journal of Development Studies* 31 (October 1994): 44-81.

de Heraclito Lima, Jose Guilherme. *Restructuring the U.S. Steel Industry: Semi-Finished Steel Imports, International Integration, and U.S. Adaptation*. Boulder: Westview, 1991.

Dick, Andrew. "Does Import Protection Act as Export Promotion? Evidence from the United States." In *Structural Change, Industrial Location, and Competitiveness*, ed. Joanne E. Oxley and Bernard Yeung, 494-512. Cheltenham: Elgar, 1998.

Dicken, Peter. *Global Shift: Reshaping the Global Economic Map in the 21st Century*, 4th ed. New York: Guilford, 2003.

Dungan, Peter. *Rock Solid: The Impact of the Mining and Primary Metals Industries on the Canadian Economy*. Toronto: Institute for Policy Analysis, 1997.

Dunn, M.L. *Recent Trends in the Seaborne Transportation of Iron and Steelmaking Raw Materials*. Brussels: International Iron and Steel Institute, 1992.

Dutz, Mark. "Economic Impact of Canadian Antidumping Law." In *Brookings Trade Forum 1998*, ed. Robert Z. Lawrence, 99-123. Washington: Brookings Institution, 1998.

Eaton, B. Curtis. "The Elementary Economics of Social Dilemmas." *Canadian Journal of Economics* 37 (November 2004): 805-29.

Eaton, Jonathan, and Edward Mirman. "Predatory Dumping and Signal Jamming." In *Trade, Policy, and International Adjustments*, ed. Akira Takayama, Hiroshi Ohta, and Michihiro Ohyama, 60-76. San Diego: Academic Press, 1991.

Eberle, Anthon, W. Kepplinger, and Siegfried Zeller. "VAI Technologies for Scrap Substitutes." In *The Steel Industry in the New Millennium, Vol. I: Technology and the Market*, ed, Ruggero Ranieri and Jonathan Aylen, 113-23. London: IOM Communications, 1998.

Eichengreen, Barry. "Dental Hygiene and Nuclear War: How International Relations Looks from Economics." *International Organization* 52 (Autumn 1998): 993-1012.

Ethier, Wilfrid. "Dumping." *Journal of Political Economy* 90 (December 1982): 487-506.

Federal Reserve Bank of Cleveland. "The Steel Industry: Economic Trends." Cleveland: April 2000.

–. "The Steel Industry: Economic Trends," Cleveland: March 2001.

Feng, Lintong. "China's Steel Industry: Its Rapid Expansion and Influence on the International Steel Industry." *Resources Policy* 20 (December 1994): 219-34.

Fischer, Bernhard, Juan-Carlos Herken-Krauer, Matthias Lucke, and Peter Nunnenkamp. *Capital-Intensive Industries in Newly Industrializing Countries: The Case of the Brazilian Automobile and Steel Industries*. Tübingen: Mohr Siebeck, 1988.

Fisher, Douglas Alan. *Steel: From the Iron Age to the Space Age*. New York: Harper and Row, 1967.

Frieden, Jeffrey A., and Ronald Rogowski. "The Impact of the International Economy on National Politics: An Analytical Overview." In *Internationalization and Domestic Politics*, ed. Robert O. Keohane and Helen V. Milner, 25-47. New York: Cambridge University Press, 1996.

Gallarotti, Giulio M. "Toward a Business-Cycle Model of Tariffs." *International Organization* 39 (Winter 1985): 155-87.

Galloway, Michael D., Bruce Blonigen, and Joseph E. Flynn. "Welfare Costs of U.S. Antidumping and Countervailing Duty Laws." *Journal of International Economics* 19 (December 1999): 211-44.

GAO (Government Accountability Office). *Foreign Investment: Laws and Policies Regulating Foreign Investment in 10 Countries*. Report 08-320 (February 2008). Washington.

–. *International Monetary Fund: Trade Policies of IMF Borrowers*. Report B-282825 (June 1999). Washington.

–. *International Trade: The Health of the U.S. Steel Industry*. Report B-236037 (July 1989). Washington.

Garrett, Geoffrey, and Peter Lange. "Internationalization, Institutions, and Political Change." In *Internationalization and Domestic Politics*, ed. Robert O. Keohane and Helen V. Milner, 79-107. New York: Cambridge University Press, 1996.

Ghemawat, Pankaj. *Games Businesses Play: Cases and Models.* Cambridge, MA: MIT Press, 1997.

Gilpin, Robert. *The Challenge of Global Capitalism: The World Economy in the 21st Century.* Princeton: Princeton University Press, 2000.

–. *Global Political Economy: Understanding the Global Economic Order.* Princeton: Princeton University Press, 2001.

–. *The Political Economy of International Relations,* Princeton: Princeton University Press, 1987.

Glais, Michel. "Steel Industry." In *European Policies on Competition, Trade, and Industry: Conflict and Complementarities,* ed. Pierre Buiges, Alexis Jacquemin, and Andre Sapir, 219-59. Aldershot: Elgar, 1995.

Gold, Bela, William S. Pierce, Gerhard Rosegger, and Mark Perlman. *Technological Progress and Industrial Leadership: The Growth of the U.S. Steel Industry, 1900-1970.* Lexington: Lexington, 1984.

Goldstein, Judith. "The Political Economy of Trade: Institutions of Protection." *American Political Science Review* 80 (March 1986): 161-84.

Greenhut, Melvin L., George Norman, and Chao-Shung Hung. *The Economics of Imperfect Competition.* Cambridge: Cambridge University Press, 1987.

Grossman, Gene M. "Strategic Export Promotion: A Critique." In *Strategic Trade Policy and the New International Economics,* ed. Paul R. Krugman, 47-68. Cambridge, MA: MIT Press, 1986.

Gupta, Poonan. "Why Do Firms Pay Antidumping Duty?" Working Paper. Washington: IMF Research Department, 1999.

Gurov, Alexander. "Steel in the Russian Federation." Geneva: International Labour Organization, September 2000.

Harrigan, James. "The Impact of the Asia Crisis on U.S. Industry: An Almost-Free Lunch?" *Federal Reserve Bank of New York Economic Policy Review* 6 (September 2000): 71-88.

Harris, Richard G. "Globalization, Trade, and Income." *Canadian Journal of Economics* 26 (November 1993): 755-76.

–. "Trade and Industrial Policy for a 'Declining' Industry: The Case of the U.S. Steel Industry." In *Empirical Studies of Strategic Trade Policy,* ed. Paul R. Krugman and Alasdair Smith, 131-56. Chicago: University of Chicago Press, 1984.

Hart, Michael. *A Trading Nation: Canadian Trade Policy from Colonialism to Globalization.* Vancouver: UBC Press, 2002.

Hartigan, James. "Predatory Dumping." *Canadian Journal of Economics* 29 (February 1996): 228-39.

Hartigan, James, Sreenivas Kamma, and Phillip Perry. "Are Subsidies More Dangerous Than Dumping? Evidence from Wealth Effects for the Steel Industry." *Journal of Economic Integration* 9 (March 1994): 45-61.

Hathaway, Oona. "Positive Feedback: The Impact of Trade Liberalization on Industry Demands for Protection." *International Organization* 52 (Summer 1998): 575-612.

Hekman, John S. "An Analysis of the Changing Location of Iron and Steel Production in the Twentieth Century." *American Economic Review* 68 (March 1978): 123-33.

Helliwell, John F. *Globalization and Well-Being.* Vancouver: UBC Press, 2002.

Hirschfeld, Julie. "Bush's Trade Agenda Starts Out Between Big Steel and a Hard Place." *Congressional Quarterly Weekly* 59 (February 24, 2001): 431-34.

Hirschman, Albert O. "The Political Economy of Import-Substitution Industrialization in Latin America." *Quarterly Journal of Economics* 82 (February 1968): 2-8.

Hiscox, Michael J. "Commerce, Coalitions, and Factor Mobility: Evidence from Congressional Votes on Trade Legislation." *American Political Science Review* 96 (September 2002): 593-608.

Hoekman, Bernard, and Michel M. Kostecki. *The Political Economy of the World Trading System: The WTO and Beyond.* Oxford: Oxford University Press, 2001.

Hogan, William. "Prospects for the Steel Industry into the New Millennium." In *The Steel Industry in the New Millennium, Vol. I: Technology and the Market,* ed. Ruggero Ranieri and Jonathan Aylen, 321-27. London: IOM Communications, 1998.

Hu, Albert G., Gary Jefferson, and Qian Jinchang. "R and D and Technology Transfer: Firm-Level Evidence from Chinese Industry." *Review of Economics and Statistics* 87 (November 2005): 780-86.

Hufbauer, Gary Clyde, and Ben Goodrich. "Steel: Big Problems, Better Solutions." Policy Brief 01-9. Washington: Institute for International Economics, July 2001.

–. "Steel Policy: The Good, the Bad, and the Ugly." Policy Brief PB03-1. Washington: Institute for International Economics, January 2003.

Hutton, Susan, and Michael Trebilcock. "An Empirical Study of the Application of Canadian Anti-Dumping Laws: A Search for Normative Rationales." *Journal of World Trade* 24 (June 1990): 123-46.

IC (Industry Canada). *Primary Steel in Canada: Industry Snapshot.* Ottawa: 2000.

–. *Primary Steel: Overview and Prospects.* Ottawa: 1996.

International Iron and Steel Institute (IISI). *World Steel in Figures.* Brussels: 2000, 2001, 2002, 2003, 2004, 2005, 2006, 2007, 2008. Annual editions.

Inwood, Kris. "The Iron and Steel Industry." In *Progress Without Planning: The Economic History of Ontario from Confederation to the Second World War,* ed. Ian M. Drummond, 185-207. Toronto: University of Toronto Press, 1987.

Jefferson, Gary H. "China's Iron and Steel Industry." *Journal of Development Economics* 33 (October 1990): 329-55.

Jones, Kent. "Structural Adjustment in the United States Steel Industry." In *Industry on the Move: Causes and Consequences of International Relocation in the Manufacturing Industry,* ed. Gijsbert van Liemt, 181-207. Geneva: International Labour Organization, 1992.

Juhn, Sung-il. "Challenge of a Latecomer: The Case of the Korean Steel Industry with Specific Reference to POSCO." In *Changing Patterns of International Rivalry: Some Lessons from the Steel Industry,* ed. Etsuo Abe and Yoshitaka Suzuki, 269-93. Tokyo: University of Tokyo Press, 1991.

Jung, Kyung-Hee. "Task Shift of Automakers to Steel Suppliers in the Value Chain of Automotive Sheets." *International Journal of Automotive Technology and Management* 5 (2005): 216-33.

Junius, Karsten. *The Economic Geography of Production, Trade, and Development.* Tübingen: Mohr Siebeck, 1999.

Kang, Nam-Hoon, and Kentaro Sakai. *New Patterns of Industrial Globalization: Cross-Border Mergers and Acquisitions and Strategic Alliances.* Paris: OECD, 2001.

Kapstein, Ethan. "Winners and Losers in the Global Economy." *International Organization* 54 (Spring 2000): 359-84.

Katzenstein, Peter, Robert O. Keohane, and Stephen D. Krasner. "*International Organization* and the Study of World Politics." *International Organization* 52 (Autumn 1998): 645-85.

Keeling, Bernard. "Structural Change in the World Steel Industry: A North-South Perspective." In *Industry on the Move: Causes and Consequences of International Relocation in the Manufacturing Industry,* ed. Gijsbert van Liemt, 149-78. Geneva: International Labour Organization, 1992.

Kelly, Kenneth H., and Morris E. Morkre. "Do Unfairly Traded Imports Injure Domestic Industries?" *Review of International Economics* 6 (May 1998): 321-32.

Kierzkowski, Henryk. "Recent Advances in International Trade Theory: A Selective Survey." *Oxford Review of Economic Policy* 3 (Spring 1987): 1-19.

Kilbourn, William. *The Elements Combined: A History of the Steel Company of Canada.* Toronto: Clarke, Irwin, 1960.

Kipping, Matthias. "Cooperation Between Steel Producers and Steel Users: A Major Determinant of National Competitive Advantage." In *The Steel Industry in the New Millennium, Vol. I: Technology and the Market,* ed. Ruggero Ranieri and Jonathan Aylen, 215-34. London: IOM Communications, 1998.

Klemperer, Paul, and Margaret Meyer. "Price Competition vs. Quantity Competition: The Role of Uncertainty." *Rand Journal of Economics* 17 (Winter 1986): 618-38.

Krugman, Paul R. *Development, Geography, and Economic Theory.* Cambridge, MA: MIT Press, 1995.

–. *Geography and Trade*. Leuven: Leuven University Press and Cambridge, MA: MIT Press, 1991.

–. "Increasing Returns, Imperfect Competition, and the Positive Theory of International Trade." In *Handbook of International Economics, Vol. 3*, ed. Gene M. Grossman and Kenneth Rogoff, 1243-77. Amsterdam: North Holland, 1995.

–. "Increasing Returns, Monopolistic Competition, and International Trade." *Journal of International Economics* 9 (November 1979): 469-79.

–. "Industrial Organization and International Trade." In *Handbook of Industrial Organization, Vol. II*, ed. Richard Schmalensee and Robert D. Willig, 1179-1223. Amsterdam: North Holland, 1989.

–. "A Model of Innovation, Technology Transfer, and World Distribution of Income." *Journal of Political Economy* 87 (April 1979): 253-66.

–. "Scale Economies, Product Differentiation, and the Pattern of Trade." *American Economic Review* 70 (December 1980): 950-59.

–. "Where Is the 'New Economic Geography'?" In *The Oxford Handbook of Economic Geography*, ed. Gordon L. Clark, Maryann P. Feldman, and Meric S. Gertler, 49-60. Oxford: Oxford University Press, 2000.

Krupp, Corinne, and Susan Skeath. "Evidence on Upstream and Downstream Impacts of Antidumping Cases." *North American Journal of Economics and Finance* 13 (August 2002): 163-78.

Kurth, James. "The Political Consequences of the Product Cycle." *International Organization* 33 (Winter 1979): 1-34.

Lacombe, Barry. "Steel: Unions, Industry, Governments, and Tribunals." *Canada–United States Law Journal* 30 (2004): 211-20.

Lankford, William T., Jr., et al., eds. *The Making, Shaping, and Treating of Steel*, 10th ed. Pittsburgh: Association for Iron and Steel Technology, 1985.

Leland, Hayne E. "Theory of the Firm Facing Uncertain Demand." *American Economic Review* 62 (June 1972): 278-91.

Liebman, Benjamin H. "Safeguards, China, and the Price of Steel." *Review of World Economics* 142 (July 2006): 354-73.

Lim, Chin. "The Ranking of Behavioural Modes of the Firm Facing Uncertain Demand." *American Economic Review* 70 (1982): 217-24.

Lindsey, Brink. "The U.S. Antidumping Law: Rhetoric Versus Reality." *Journal of World Trade* 34 (January 2000): 1-38.

Lungen, Hans Bodo. "Technological Innovations in Iron and Steelmaking and Their Effects on Coal, Coke, and Iron Ores." In *The Steel Industry in the New Millennium, Vol. I: Technology and the Market*, ed. Ruggero Ranieri and Jonathan Aylen, 39-59. London: IOM Communications, 1998.

Lydolph, Paul E. *Geography of the U.S.S.R.* New York: Wiley, 1970.

Maddala, G.S., and Peter T. Knight. "International Diffusion of Technical Change: A Case Study of the Oxygen Steelmaking Process." *Economic Journal* 77 (September 1967): 531-58.

Mankiw, N. Gregory, and Michael D. Whinston. "Free Entry and Social Inefficiency." *Rand Journal of Economics* 17 (Spring 1986): 48-58.

Marcus, Peter F. "Ironies in Steel." In *The Steel Industry in the New Millennium, Vol. I: Technology and the Market*, ed. Ruggero Ranieri and Jonathan Aylen, 7-31. London: IOM Communications, 1998.

Markusen, James R. "Comment" [to Michael O. Moore, "The Waning Influence of Big Steel?"] in *The Political Economy of American Trade Policy*, ed. Anne O. Krueger, 127-30. Chicago: University of Chicago Press, 1996.

Martins, Joachim Oliveira. "Market Structure, Trade, and Industry Wages." OECD Economic Studies no. 22. Paris: Spring 1994.

Masi, Anthony. "Structural Adjustment and Technological Change in the Canadian Steel Industry, 1970-1986." In *The New Era of Global Competition: State Policy and Market Power*, ed. Daniel Drache and Meric S. Gertler, 181-205. Montreal and Kingston: McGill-Queen's University Press, 1991.

Mastel, Greg. "The U.S. Steel Industry and Antidumping Law." *Challenge* 42 (May-June 1999): 84-94.

Matsushita, Mitsuo, Thomas J. Schoenbaum, and Petros C. Mavroidis. *The World Trade Organization: Law, Practice, and Policy.* Oxford: Oxford University Press, 2003.

McKeown, Timothy J. "Firms and Tariff Regime Change: Explaining the Demand for Protection." *World Politics* 36 (January 1984): 215-33.

Messerlin, Patrick A. ""China in the World Trade Organization: Antidumping and Safeguards." In *The WTO After Seattle,* ed. Jeffrey J. Schott, 159-83. Washington: Institute for International Economics, 2000.

Metals and Minerals Processing Branch. *Primary Steel: Overview and Prospects.* Ottawa: Minister of Supply and Services, 1996.

Miller, Robert R. "The Changing Economics of Steel." *Finance and Development* 28 (June 1991): 38-40.

Milner, Helen V. *Interests, Institutions, and Information: Domestic Politics and International Relations.* Princeton: Princeton University Press, 1997.

–. "Resisting the Protectionist Temptation: Industry and the Making of Trade Policy in France and the United States in the 1970s." *International Organization* 41 (Fall 1987): 639-65.

Miranda, Jorge, Raul A. Torres, and Mario Ruiz. "The International Use of Antidumping: 1987-1997. *Journal of World Trade* 32 (October 1998): 5-71.

Montero, Alfred P. "State Interests and the New Industrial Policy in Brazil: The Privatization of Steel, 1990-1994." *Journal of Interamerican Studies and World Affairs* 40 (Fall 1998): 27-63.

Moore, Michael O. "Made in America? U.S. Steelmaking in the 1990s: An American Resurgence?" *International Executive,* 38 (July-August 1996): 431-63.

–. "Steel Protection in the 1980s: The Waning Influence of Big Steel?" Working Paper no. 4760. Cambridge, MA: National Bureau of Economic Research, June 1994.

–. "Waning Influence of Big Steel?" In *The Political Economy of American Trade Policy,* ed. Anne O. Krueger, 73-125. Chicago: University of Chicago Press, 1996.

Moreau, Alain. "Intermaterial Competition Beyond Technological and Economic Factors." In *The Steel Industry in the New Millennium, Vol. I: Technology and the Market,* ed. Ruggero Ranieri and Jonathan Aylen, 263-69. London: IOM Communications, 1998.

Mueller, Hans G. "The Steel Industry." *Annals of the American Academy of Political and Social Science* 460 (March 1982): 73-82.

Neill, Jon R. "Production and Production Functions: Some Implications of a Refinement to Process Analysis." *Journal of Economic Behaviour and Organization* 51 (August 2003): 507-21.

Nicolaides, Phaedon. "The Competition Effects of Dumping." *Journal of World Trade* 24 (October 1990): 115-31.

Niels, Gunnar. "What Is Antidumping Policy Really About?" *Journal of Economic Surveys* 14 (September 2000): 467-92.

OECD, *Predatory Pricing,* Paris, 1989.

OECD Business and Industry Policy Forum. Forum on Global Industrial Restructuring. Report: "Trends in Global Industrial Restructuring," Paris: February 19, 2002.

OECD Directorate for Financial, Fiscal, and Enterprise Affairs. Committee on Competition Law and Policy. *Oligopoly,* Paris: October 19, 1999.

OECD Directorate for Science, Technology, and Industry. "The Outlook for Steel." Background and Issues Paper for OECD Special Meeting at High Level on Steel Issues. Paris: January 12-13, 2005.

Ohashi, Hiroshi. "Learning by Doing, Export Subsidies, and Industry Growth: Japanese Steel in the 1950s and 1960s." *Journal of International Economics* 66 (July 2005): 297-323.

O'hUallachain, Breandán. "The Restructuring of the U.S. Steel Industry: Changes in the Location of Production and Employment." *Environment and Planning A* 26 (September 1993): 1339-59.

Park, Donghyun. "Why Do Firms Dump at a Loss? An Economies-of-Scale Explanation." *Open Economies Review* 9 (1998): 257-62.

Park, Kwang Soo, and MoonJoong Tcha. "The Korean Steel Industry After the Economic Crisis: Challenges and Opportunities." In *The Korean Economy at the Crossroads*, ed. Moon-Joong Tcha and Chung-Sok Suh, 200-14. London: RoutledgeCurzon, 2003.

Philips, Louis. *The Economics of Price Discrimination*. Cambridge: Cambridge University Press, 1983.

Prusa, Thomas. "An Overview of the Impact of U.S. Unfair Trade Laws." In *Trading Punches: Trade Remedy Law and Disputes Under NAFTA*, ed. Beatriz Leycegni, William B.P. Robson, and S. Dahlia Stein, 183-205. Washington: National Planning Association, 1995.

Rees, Ray. "Tacit Collusion." *Oxford Review of Economic Policy* 9 (Summer 1993): 27-40.

Repetto, Eugenio, Gustavo Brascugli, and Giovanni Perni. "Evolution of EAF Steelmaking Route." In *The Steel Industry in the New Millennium, Vol. I: Technology and the Market*, ed. Ruggero Ranieri and Jonathan Aylen, 125-55. London: IOM Communications, 1998.

Richardson, J. David. "The New Political Economy of Trade Policy." In *Strategic Trade Policy and the New International Economics*, ed. Paul R. Krugman, 257-82. Cambridge, MA: MIT Press, 1986.

Robson, William B.P., S. Dahlia Stein, and Rafael Fernandez de Castro. "What's the Fight About? An Overview of Trade Disputes in North America." In *Trading Punches: Trade Remedy Law and Disputes Under NAFTA*, ed. Beatriz Leycegni, William B.P. Robson, and S. Dahlia Stein, 1-23. Washington: National Planning Association, 1995.

Rogowski, Ronald. "Trade and the Variety of Democratic Institutions." *International Organization* 41 (Spring 1987): 203-23.

Rubin, Jeffrey, and Benjamin Tal. "Will Soaring Transport Costs Reverse Globalization?" *StrategEcon*, CIBC World Markets Inc., May 27, 2008, 4-7.

Sabry, Faten. "An Analysis of the Decision to File, the Dumping Estimates, and the Outcome of Antidumping Petitions." *International Trade Journal* 14 (Summer 2000): 109-45.

Sagers, Matthew. "The Iron and Steel Industry in Russia and the CIS in the Mid-1990s." *Post-Soviet Geography and Economics* 37 (1996): 195-263.

Sato, Hajime. "Total Factor Productivity vs. 'Realism': The Case of the South Korean Steel Industry." *Cambridge Journal of Economics* 29 (July 2005): 635-55.

Schelling, Thomas C. *The Strategy of Conflict*. New York: Oxford University Press, 1963.

Schmitz, James A., Jr. "What Determines Productivity? Lessons from the Dramatic Recovery of the U.S. and Canadian Iron Industries Following Their Early 1980s Crisis." *Journal of Political Economy* 113 (June 2005): 582-625.

Schwanen, Daniel. "When Push Comes to Shove: Quantifying the Continuing Use of Trade 'Remedy' Laws Between Canada and the United States." In *Trading Punches: Trade Remedy Law and Disputes Under NAFTA*, ed. Beatriz Leycegni, William B.P. Robson, and S. Dahlia Stein, 161-82. Washington: National Planning Association, 1995.

Schwartz, Herman M. *States Versus Markets: History, Geography, and the Development of the International Political Economy*. New York: St. Martin's, 1994.

Schwartz, Nelson. "Bent but Unbowed." *Fortune*, July 22, 2002, 118.

Shapiro, Carl. "Theories of Oligopoly Behaviour." In *Handbook of Industrial Organization, Vol. I*, ed. Richard Schmalensee and Robert D. Willig, 329-414. Amsterdam: North-Holland, 1989.

Shin, Hyun Ja. "Possible Instances of Predatory Pricing in Recent U.S. Antidumping Cases." In *Brookings Trade Forum 1998*, ed. Robert Z. Lawrence, 81-97. Washington: Brookings Institution, 1998.

Shin, Jang-Sup. *The Economics of the Latecomers: Catching Up, Technology Transfer, and Institutions in Germany, Japan, and South Korea*, London: Routledge, 1996.

Shinnan, Andy, and Kristina Shimmons. "Primary Metal Industries." Statistics Canada, Manufacturing, Construction, and Energy Division, April 2000.

Siekman, Philip. "Good Steel Made Cheaply." *Fortune* 147, May 12, 2003, 144B.

Staiger, Robert W., and Frank Wolak. "The Effect of Domestic Antidumping Law in the Presence of Foreign Monopoly." *Journal of International Economics* 32 (May 1992): 265-87.

Steele, Peter J.B. "The Propensity of Advanced Free World Economies to Import Steel." *Canada–United States Law Journal* 2 (1979) 17-38.

"Steeling for a Fight." *Time Atlantic,* March 18, 2002, 38.

Stegemann, Klaus, "Policy Rivalry Among Industrial States: What Can We Learn from Models of Strategic Trade Policy?" *International Organization* 41 (Winter 1989): 73-100.

Stewart, Frances, "Recent Theories of International Trade: Some Implications for the South." In *Monopolistic Competition and International Trade,* Henryk Kierzkowski, 84-108. Oxford: Clarendon, 1984.

Stiglitz, Joseph. "Dumping on Free Trade: The U.S. Import Trade Laws." *Southern Economic Journal* 64 (October 1997): 402-24.

Stocking, George W. *Basing Point Pricing and Regional Development: A Case Study of the Iron and Steel Industry.* Chapel Hill: University of North Carolina Press, 1954.

Stokes, Bruce. "Steel Thyself, America." *National Journal,* March 9, 2002, 717.

Storper, Michael. "Globalization, Localization, and Trade." In *The Oxford Handbook of Economic Geography,* ed. Gordon L. Clark, Maryann Feldman, and Meric S. Gertler, 146-65. Oxford: Oxford University Press, 2000.

–. "Territories, Flows, and Hierarchies in the Global Economy." In *Reading Economic Geography,* ed. Trevor J. Barnes, Jamie Peck, Eric Sheppard, and Adam Ticknell, 271-89. Malden: Blackwell, 2004.

Strange, Susan. "The Management of Surplus Capacity." *International Organization* 33 (Summer 1979): 303-34.

Sun, Pei. "Industrial Policy, Corporate Governance, and the Competitiveness of China's National Champions: The Case of Shanghai Baosteel Group." *Journal of Chinese Economic and Business Studies* 3 (May 2005): 173-92.

–. "Is the State-Led Industrial Restructuring Effective in Transition China? Evidence from the Steel Sector." *Cambridge Journal of Economics* 31 (July 2007): 601-24.

Suzuki, Yoshitaka. "The Rise and Decline of Steel Industries: A Business Historical Introduction." In *Changing Patterns of International Rivalry: Some Lessons from the Steel Industry,* ed. Etsuo Abe and Yoshitaka Suzuki, 1-17. Tokyo: University of Tokyo Press, 1991.

Swagel, Philip. *Union Behaviour, Industry Rents, and Optimal Policies.* Washington: International Monetary Fund, 1996.

Sykes, Alan O. "Antidumping and Antitrust: What Problems Does Each Address?" In *Brookings Trade Forum 1998,* ed. Robert Z. Lawrence, 1-43. Washington: Brookings Institution, 1998.

Tarr, David G. "Cyclical Dumping: The Case of Steel Products." *Journal of International Economics* 9 (January 1979): 57-63.

–. "The Minimum Efficient Size Steel Plant." *Atlantic Economic Journal* 12 (March 1984): 122.

Tcha, Moonjoong, and Larry A. Sjaastad. "Analysis of Steel Prices." In *The Economics of the East Asia Steel Industries,* ed. Yanrui Wu, 207-24. Aldershot: Ashgate, 1998.

Thomas, Jeffrey S., and Michael Meyer. *The New Rules of Global Trade: A Guide to the World Trade Organization.* Scarborough: Carswell, 1997.

Thomson, William R., and Rafael Reuveny. "Tariffs and Trade Fluctuations: Does Protectionism Matter as Much as We Think?" *International Organization* 52 (Spring 1998): 421-40.

Tivig, Thusnelda, and Uwe Waltz. "Market Share, Cost-Based Dumping, and Anti-Dumping Policy." *Canadian Journal of Economics* 33 (February 2000): 69-86.

Tornell, Aaron. "Rational Atrophy: The U.S. Steel Industry." Working Paper 6084. Cambridge, MA: National Bureau of Economic Research, 1997.

Truett, Lila J., and Dale B. Truett. "The Korean Metals Industry and Economic Development." *Journal of Asian Economics* 3 (Summer 1997): 333-47.

Valenti, Michael. "Vacuum Degassing Yields Stronger Steel." *Mechanical Engineering,* April 1998, 54-58.

Vandenbussche, Hylke, and Xavier Wauthy. "Inflicting Injury Through Product Quality: How European Antidumping Policy Disadvantages European Producers." *European Journal of Political Economy* 17 (March 2001): 101-16.

Varian, Hal R. "Price Discrimination." In *Handbook of Industrial Organization, Vol. I,* ed. Richard Schmalensee and Robert D. Willig, 597-654. Amsterdam: North Holland, 1989.

Venables, Anthony. "Equilibrium Locations of Vertically Linked Industries." *International Economic Review* 37 (May 1996): 341-59.

Vernon, Raymond. *Sovereignty at Bay.* New York: Basic, 1971.

Viner, Jacob. *Dumping: A Problem in International Trade.* New York: August M. Kelly, 1966.

Warren, Kenneth. *World Steel: An Economic Geography.* New York: Crane, Russak, 1975.

Warrian, Peter, and Celine Mulhern. "Knowledge and Innovation in the Interface Between the Steel and Automotive Industries: The Case of Dofasco." *Regional Studies* 39 (April 2005): 161-70.

Watson, Thomas. "Heavy Metal." *Canadian Business,* November 21, 2005, 35-36.

–. "What a Mess." *Canadian Business,* January 30, 2006, 36-44.

Webb, Michael. "Ambiguous Consequences of Anti-Dumping Laws." *Economic Inquiry* 30 (July 1992): 437-49.

Webber, Michael. "Regional Production and the Production of Regions: The Case of Steeltown." In *Production, Work, Territory: The Geographical Anatomy of Industrial Capitalism,* ed. Allan J. Scott and Michael Storper, 197-224. Boston: Allen and Unwin, 1986.

Weinstein, David. "Competition and Unilateral Dumping." *Journal of International Economics* 32 (May 1992): 379-88.

Willig, Robert D. "Economic Effects of Antidumping Policy." In *Brookings Trade Forum 1998,* ed. Robert Z. Lawrence, 57-79. Washington: Brookings Institution, 1998.

Wood, Adrian. *North-South Trade, Employment, and Inequality: Changing Fortunes in a Skill-Driven World.* Oxford: Clarendon, 1994.

Woolcock, Stephen. "Iron and Steel." In *The International Politics of Surplus Capacity,* ed. Susan Strange and Roger Tooze, 69-79. London: George Allen and Unwin, 1981.

Wu, Yanrui. "China's Metals Industry (I)." In *The Economics of the East Asia Steel Industries,* ed. Yanrui Wu, 85-115. Aldershot: Ashgate, 1998.

News and Reportage

Advanced Materials and Processes
Automotive Body Repair News
Automotive Design and Production
Automotive News
Business Week
Businessline
China Daily
Cleveland Plain Dealer
Design News
The Economist
Financial Executive
Financial Post
Financial Times
Globe and Mail
The Guardian
Industry Week
International Herald Tribune
International Trade Finance
Iron Age New Steel
Jiji Press English News Service
Journal of Commerce
Manufacturing Engineering
Mechanical Engineering
Metal Bulletin Monthly
Metal Center News
Metal Producing and Processing
Natural Gas Week
New Steel
New York Times
Northern Miner
Ottawa Citizen

Platt's International Coal Report
Portland Business Journal
Purchasing
St. Catharines Standard
Wall Street Journal
Ward's Auto World

Index

Aceralia Corporacion Siderurgica SA, 57
Acme Steel, 57
AK Steel, 55, 154
Alcan Inc., 3
Algoma Steel, 2, 13, 30, 53, 55, 72
American Iron and Steel Institute, 63
anti-dumping rules: critique, 14; frequency
    of use, 143; justification, 11-12, 68, 144-
    45; long-term protection, 143; negative
    effects, 148-50; pattern of protection,
    Canada, US, 145; predation, 147-48
Arbed SA, 57
Arcelor, 2; acquisition of Dofasco, 5,
    122-23, 154; expansion in Brazil, 109;
    in China, 157; US joint venture, 93
ArcelorMittal Dofasco, 4, 29
ArcelorMittal Steel: automobile industry,
    8, 165; world rank, 103
Atlas Steels, 27
automotive industry: alliances with steel-
    makers, 11, 165-69; demand for advanced
    steels, 27, 44-45, 161, 164; interorgan-
    izational structure, 163; Southern
    United States, 93; steel fabrication, 167;
    steel mergers and acquisitions, 4, 8,
    122-23; steel supply patterns, 162

Baosteel, 6, 20, 31, 94, 116, 182; joint
    venture with Arcelor and Nippon, 122;
    joint venture in Brazil, 108, 122
Barnett, Donald, 120
basic oxygen furnace: adoption, 23-24;
    chemistry, 22; cost comparisons, 23
Belassa, Bela, 78
Bessemer converter, 1; chemistry, produc-
    tion costs, 22
Bethlehem Steel Corp., 2, 57, 123, 136
BHP Billiton PLC, 30, 156

blast furnace: China, steel industry, 20;
    economies of scale, 20-21; future, 41;
    production costs, 18; steelmaking
    process, chemistry, 18
Brander, James, 66
Brazilian steel industry: advantages,
    107, 121; expansion into United States,
    109-10; joint steelmaking ventures, 20,
    108-9; state involvement, 105-6
British Steel, 4, 57, 184

Canada-United States steel trade, 33, 47,
    78, 84, 89, 91, 93, 129, 136
Canadian Steel Producers Association,
    63
Chapparal Steel, 13
China: iron ore imports, 94, 117; pro-
    spective surpluses, 52, 140-41; rapid
    growth of demand, 87; rapid steel
    industry expansion, 88, 116-18; state
    and steel industry, 43, 116; steel
    industry structure, 116-17
Cleveland Cliffs, Inc., 20
Cliffs Natural Resources. *See* Cleveland
    Cliffs, Inc.
coal: pricing, 32-33
collusion, 63-64
Companhia Vale do Rio Doce (CVRD), 3,
    106
comparative advantage: explanation of
    trade, 83, 89
competition law: consolidation, 8; market
    apportionment, 47
Competition Policy Review Panel, 6, 189
concentration: steel industry, 142, 153
consolidation, 3-4; acquisition from
    periphery, 13, 180-85; ArcelorMittal as
    trendsetter, 152; Canada, 153; China,

116-17; competition law, 172-73; evaluated, 170-71; excess capacity, 160; motives, benefits, limits, 10-11, 52, 154-57; price leadership, 46-47, 155; pricing power, 158; trade policy preferences, 78; United States, 153

continuous casting, 26-28; Brazil, 106; Canada, United States, 56; China, 116; Russia, 113; use worldwide, 28

Corex process, 20

Corus Steel, 4, 13, 32, 114, 165, 184

costs of trade: acquisition of foreign steelmakers, steel export volume, 7; future of world steel industry, 14, 177-79

Cournot competition, 66

CSN Steel (Companhia Siderurgica Nacional), 106, 156

CST Steel (Companhia Siderurgica de Tubarao), 106, 156

Dofasco, 3, 4, 23, 53, 55, 56; US joint venture, 93. *See also* ArcelorMittal Dofasco

dumping, 11, 65, 67-68; competition law, predation, 75-76; discounting strategies, 71; intentional and innocent, 69-70; trade policy preferences, 78; welfare effects, 75. *See also* trade protection

economic geography: centre-periphery, 13, 92; explanation of trade, 12, 14, 83, 90-93, 100; industrial location, 12, 90-92, 97-98; international division of labour, 12, 127-30; reversal, 185-88; role of the state, 12, 90, 100-1; transferable technology, 97-98

economies of scale: Canada, 53; China, 116; competition, 71; consolidation, 155; free trade, 74; industrial location, 90-92; origin of trade, 90, 100-1; steelmaking, 42-43; United States, 54. *See also* economic geography, explanation of trade

electric furnace: steelmaking, minimills, 24-25

Essar Global Ltd., 4, 13, 30; purchase of Esmark, Inc., 114. *See also* Essar Steel Algoma

Essar Steel Algoma, 4, 114, 184

Eurofer, 87

European Commission, 6

European Union (EU), 5

Evraz, 31, 93, 113; international expansion, 114-15

exports: market penetration, 71

factor mobility, 43, 46; protectionism, 12, 74, 77

foreign acquisitions: national security, Canada, Germany, United States, Japan, 6

Foreign Investment National Security Act, 6, 189

General Agreement on Tariffs and Trade, 2. *See also* World Trade Organization

General Motors Corp., 8

Gerdau Ameristeel, 13

Gerdau SA, 106, 129

Gilpin, Robert, 123, 129

globalization: future of steel industry, 175-80; reversible, 185-87; sources, 4-5

Goodrich, Ben, 46, 139

Hanbo Steel, 105

Heartland Steel: purchase by CSN, 156

Helliwell, John, 178

high manganese twinning-induced plasticity steel (TWIP), 165

high-strength and ductility steel (HSD), 165

Hufbauer, Gary Clyde, 46, 139

Hyundai Motors, 20, 93, 105

import substitution, 99

Inchon Iron and Steel, 105

industrial regions: logistics, 90-91; role of the state, 100-1; world markets, 97

Inland Steel, 3, 57

International Nickel Company, 3

International Steel Group, 2, 57

international steel market: oversupply, 14

intra-industry trade, 129

Ipsco Steel, 46, 93; purchase by Evraz, 115, 189

iron ore: Australia and Brazil, 30; cost advantages, 21, 32; North America, 29-30; oligopoly, 31; transportation, remote supply, 30, 31-32

Ironmaking Technology Mark, 3, 20

Japan: export advantages, 102-3; postwar steel industry, 102

JFE Holdings, 103, 116, 165

Jones and Laughlin Steel, 57

Kaldor, Nicholas, 181-82

Katzenstein, Peter, 80

Kawasaki Steel, 103. *See also* JFE Holdings

Keohane, Robert O., 80

Kobe Steel, 20, 116

Koninklijke Hoogovens, 57
Krasner, Stephen, 80
Krivorozhstal, 32
Krugman, Paul, 66

labour costs: industrial location, modernization, 128; Russia (compared), 113
Lardy, Nicholas, 141
LTV Corporation, 57, 136

Magnitogorsk Metallurgical Combine, 31, 113; international expansion, 115
McLouth Steel, 23
minimills, 10; global production, 34-35; locational mobility, 92-93; pricing influence, 48, 65; regional markets, 35, 93; scrap, 33; substitutable products, 45; technology, costs, 33, 41. *See also* steel industry, sectoral division
Mittal, Aditya, 154, 157
Mittal, Lakshmi, 3, 10, 154
Mittal Steel, 3, 57; expansion in China, 157; merger with ISG, 98; takeover of Arcelor, 5, 159. *See also* ArcelorMittal Steel

Nash, John. *See* Nash equilibrium
Nash equilibrium, 64
National Steel, 56, 136
Nippon Steel, 3, 31, 116, 154; world rank, 103
NKK Corporation, 103. *See also* JFE Holdings
North American Free Trade Agreement (NAFTA), 78
Novolipetsk, 31, 113; international expansion, 115
Nucor Corp., 3, 56, 154

OAO Severstal, 4, 12, 31, 57, 93, 113, 156; expansion in United States, 114, 184
oligopoly, 11; collusion, 46, 64; defensive consolidation, 156; incentives, behaviour, 62-64; price leadership, 10, 63. *See also* steel industry, economic characteristics
Ontario government, 2, 58
open-hearth furnace, 22; current use, 24; Russia, 112-13
Oregon Steel, 115, 189

Pension Benefit Guaranty Corporation, 57
POSCO, 31, 116; recipe sharing, 105; technology, advantages, growth, 103-4
price discrimination, 65-69
product cycle theory, 119, 123-25

reciprocal dumping, 66
Republic Steel, 57
Rio Tinto Group, 3, 29, 30-31, 94, 156
Rogowski, Ronald, 76
Ross, Wilbur, 57, 123
Rouge Industries: purchase by OAO Severstal, 4, 58, 114, 122, 157
Rubin, Jeffrey, 186-87
Russia: expansion in United States, 114-15; export advantages, 113; Kremlin, 115, 189; Soviet era, privatization, 111-12; steel industry, decline and recovery, 112-13

SeverCorr, 93, 114
smelting reduction, 20
South Korea, 20; steel use, 87. *See also* POSCO
state: national interest, national security, 5, 188-91; role in industrial development, 99-100, 121, 125-27. *See also* economic geography
state-linked investment funds: national security, strategic industries, 15, 188-90
steel: basic cast forms, 26; carbon content, 21; export commodity, 5; imports, exports, consumption, 85-89; inelastic demand, 47-48; product differentiation, 5, 64, 98; strategic commodity, 5; substitutability, 5, 10, 44-45, 49, 64, 98, 175-76, 179
steel industry: automation, 29; bankruptcies, 136; business cycle, 47; capacity growth, 142; capital markets, 37; comparative labour costs, 119-21; concentration, 153; economic characteristics, 9-11, 36-45, 72; entry, exit barriers, 41-42; excess capacity, 72, 139-40; external economies of scale, 51; fixed costs, 36, 44; future organization, 177-80; homogenizing forces, 176; international hierarchy, 181-85; international ownership, 98; investment from periphery, 129; joint ventures, 127; marginal costs, 45-46; market access, 122; minimum efficient scale, 38-42; modernization, 37-38, 49-52, 57; new locales, 99-100; price volatility, 44, 47-48, 135-36, 140; price warfare, 43; productivity, 29; recovery, 56-58; risk, 36-37; scale economies, 42-43; sectoral division, 9, 17; specialization, 40, 51-52; surplus capacity and trade, 137-42; trade policy preferences, 79-80; technology and export advantages, 121-22; technology

transfer, 122; variable costs, output, pricing, 39
steel industry, Canada: historical development, 46-47, 53-55
steel industry, United States: historical development, 53; modernization, 56
steel scrap, 22; imports/exports, 33; prices, 34
steel service centres, 36; and automobile industry, 166; concentration, 158
steel subsidies: governments, 71-73
steelmaking: automation, 28; continuous process, 28-29; external scale economies, 51; indivisibilities, 38; minimum efficient scale, 38-39, 40
Stelco, 3, 53, 55, 57, 109, 136. *See also* United States Steel Canada
Stolper-Samuelson theorem, 119, 127
subsidies: motives, 71-72; permitted but actionable, 7, 73; welfare effects, 73
Sumitomo Steel, 116
Surma, John: president and CEO, United States Steel, 120
Svenskt Stal AB, 93, 189
Sydney Steel Corporation, 72

Tata Steel Group, 4, 13, 114, 184
technology diffusion, 123-25; joint ventures, 122, 182; transfer, industrial location, 97
ThyssenKrupp AG, 4, 93, 122-23, 165
Tokyo Steel, 102
trade protection, 11; orientation of government, 76; preferences, 76-81. *See also* dumping

transportation costs: energy prices, 95, 185-87; industrial location, 92-93, 96; industrial regions, 90-91; intermediate goods, 96; iron ore, coal, 94; overland and overseas, 92-95; trade, 90. *See also* economic geography, explanation of trade

United States: special steel tariffs 2002, 150-52
United States Steel Canada, 4, 29. *See also* Stelco
United States Steel Corp., 4, 56, 154
Usiminas (Usinas Siderurgicas de Minas Gerais), 106
Usinor SA, 57

Vale, 3, 20, 30-31, 156. *See also* Companhia Vale do Rio Doce
Value Line Investment Survey, 3
Viner, Jacob, 11-12; justifiable anti-dumping, 68, 74, 144-45, 148; typology of dumping, 67-70

Wheeling-Pittsburgh Steel Corp., 114
Whirlpool Corp., 158
Wierton Steel, 57
World Bank, 99
world steel output, 1
World Trade Organization, 2; principle of national treatment, 5; rules on dumping, 7, 65; rules on subsidies, 7, 73; safeguard rules, 7

Set in Stone by Artegraphica Design Co. Ltd.

Copy editor: Matthew Kudelka

Marquis Book Printing Inc.

Québec, Canada
2009

١